Jan Dohl

Blind Estimation and Mitigation of Nonlinear Channels

Beiträge aus der Informationstechnik

Mobile Nachrichtenübertragung

Nr. 75

Jan Dohl

Blind Estimation and Mitigation of Nonlinear Channels

VOGT

Dresden 2015

Bibliografische Information der Deutschen Nationalbibliothek
Die Deutsche Nationalbibliothek verzeichnet diese Publikation in der Deutschen Nationalbibliografie; detaillierte bibliografische Daten sind im Internet über http://dnb.dnb.de abrufbar.

Bibliographic Information published by the Deutsche Nationalbibliothek
The Deutsche Nationalbibliothek lists this publication in the Deutsche Nationalbibliografie; detailed bibliographic data are available on the Internet at http://dnb.dnb.de.

Zugl.: Dresden, Techn. Univ., Diss., 2014

Die vorliegende Arbeit stimmt mit dem Original der Dissertation „Blind Estimation and Mitigation of Nonlinear Channels“ von Jan Dohl überein.

Gesetzt vom Autor

ISBN 978-3-938860-92-2

Jörg Vogt Verlag
Niederwaldstr. 36
01277 Dresden
Germany

Phone: +49-(0)351-31403921
Telefax: +49-(0)351-31403918
e-mail: info@vogtverlag.de
Internet : www.vogtverlag.de

Technische Universität Dresden

Blind Estimation and Mitigation of Nonlinear Channels

Jan Dohl

von der Fakultät Elektrotechnik und Informationstechnik der
Technischen Universität Dresden

zur Erlangung des akademischen Grades eines

Doktoringenieurs

(Dr.-Ing.)

genehmigte Dissertation

Vorsitzender:	Prof. Dr.-Ing. habil. Renate Merker
Gutachter:	Prof. Dr.-Ing. Dr. h.c. Gerhard P. Fettweis
	Prof. Dr.-Ing. Dr.-Ing. habil. Johannes Huber

Tag der Einreichung: 01. September 2014
Tag der Verteidigung: 27. November 2014

Abstract

The worldwide demand for wireless data services is on the rise for several years now. With the introduction of smartphones and tablets, this trend has intensified. Many users now own multiple devices and data needs to be easily available across all of them. This has lead to an embrace of cloud services not only for documents, but also for photos, music and even video data, yielding another spike in traffic demand.

Multicarrier modulation is the current answer to the ever rising traffic demand. It allows the efficient usage of large bandwidths at relatively low computational complexity. It has been in use in digital video broadcasting and wireless local area networks for a while and is now introduced in cellular communications with the advent of LTE.

The downside of multicarrier modulation is its very high dynamic range which results in instantaneous power peaks. Amplifiers need to be driven with a large power reserve in order to cope with these peaks. This however reduces their energy efficiency dramatically which is especially bad for battery powered devices. Furthermore, development and production of these highly linear amplifiers is costly.

Driving the amplifiers with less reserve causes nonlinear distortion of the power peaks and hence a significant reduction in signal quality. In the past, both transmitter and receiver based methods have been presented to mitigate these distortions. However, very often perfect knowledge of the nonlinearity characteristic is assumed which is not realistic especially in receiver based methods.

The focus of this thesis is on methods that estimate the nonlinearity characteristic on the receiver side. Of special interest are so called blind algorithms because they don't require special pilot signals and hence can be used with existing standards. The main focus is on the formal derivation of blind estimation methods and their low-complexity implementation. It is shown that there is virtually no performance gap between estimated and perfect nonlinearity knowledge.

One of the methods has been implemented on a software defined radio platform. It is shown that significant performance gains can be reached for real nonlinear amplifiers. The system runs in real time on cheap off-the-shelf components proving the low complexity of the method which is ready for implementation on today's hardware.

Kurzfassung

Die weltweite Nachfrage nach drahtlosen Breitbanddiensten wächst seit mehreren Jahren kontinuierlich an. Dieser Trend hat sich mit der Einführung von Smartphones und Tablets weiter verstärkt. Viele Nutzer besitzen mehrere mobile Geräte und benötigen eine einfache Lösung zur Synchronisation. Die Cloud erfüllt dieses Bedürfnis und wird nicht nur für Dokumente, sondern auch für Fotos, Bilder und Videos benutzt. Dies treibt die Nachfrage nach Datendiensten zusätzlich an.

Die heutige Antwort liegt im Einsatz von Mehrträgermodulation. Sie verbindet die effiziente Nutzung breitbandiger Signale mit moderatem Rechenaufwand. Mehrträgermodulation ist seit mehreren Jahren im digitalen Rundfunk und drahtlosen Netzwerken im Einsatz und mit der Einführung von LTE auch im Mobilfunkbereich angekommen.

Ein Nachteil ist der hohe Dynamikumfang von Mehrträgersignalen, der sich in hohen kurzzeitigen Leistungsspitzen äußert. Um diese Spitzen nicht zu verzerren, müssen Verstärker mit hoher Leistungsreserve betrieben werden. Die Effizienz wird dadurch stark reduziert was besonders für batteriebetriebene Geräte von Nachteil ist. Außerdem sind diese hochgradig linearen Verstärker sehr teuer in Entwicklung und Produktion.

Werden die Verstärker mit weniger Leistungsreserve betrieben, wird die Signalqualität durch nichtlineare Verzerrung der Leistungsspitzen deutlich reduziert. In den letzten Jahren wurden viele Ansätze zur sender- und empfängerseitigen Korrektur nichtlinear verzerrter Signale vorgestellt. Oftmals wird in den jeweiligen Publikationen perfekte Kenntnis der Verstärkerkennlinie angenommen. Dies ist jedoch speziell für empfängerseitige Algorithmen unrealistisch.

Der Fokus dieser Dissertation liegt daher auf der empfängerbasierten Schätzung der Kennlinie der Nichtlinearität. Sogenannte blinde Algorithmen sind hierbei von besonderem Interesse weil sie keine speziellen Pilotsignale benötigen und daher kompatibel zu bestehenden Standards sind. Der Großteil der Arbeit besteht einerseits aus der formellen Herleitung blinder Schätzverfahren, andererseits aus dem Ableiten von Berechnungsverfahren mit geringer Komplexität. Es wird gezeigt, dass es praktisch keinen Unterschied zwischen der erreichbaren Systemperformance bei geschätzer und perfekter Nichtlinearitätskennlinie gibt.

Ein Algorithmus wurde auf einer Software Defined Radio Plattform implementiert. Es wird gezeigt, dass auch bei Einsatz realer Verstärker im Sättigungsbetriegb eine deutliche Performancesteigerung möglich ist. Das System ist mit preiswerten standardkomponenten umgesetzt und ermöglicht Echtzeitbetrieb, was die niedrige Komplexität der Algorithmen und damit die Einsatzfähigkeit in heutigen Systemen beweist.

Contents

List of Figures

List of Tables

List of Acronyms

3DCSI 3D Chip Stack Interconnects

ADSL Asymmetric Digital Subscriber Line

AM/AM Amplitude to Amplitude

AM/PM Amplitude to Phase

AWGN Additive White Gaussian Noise

BER Bit Error Rate

CDF Cumulative Distribution Function

CP Cyclic Prefix

CRLB Cramér-Rao Lower Bound

CSS3 Cascading Style Sheets v3

DAB Digital Audio Broadcasting

DAC Digital-to-Analog Converter

DC Direct Current

DFG Deutsche Forschungs Gesellschaft

DFR Decision Feedback Receiver

DFT Discrete time Fourier Transformation

DVB-T Terrestrial Digital Video Broadcasting

FBMC Filter Bank Multi Carrier

FFT Fast Fourier Transformation

GFDM Generalized Frequency Division Multiplexing

GUI Graphical User Interface

HAEC Highly Adaptive Energy-Efficient Computing

HTML5 Hypertext Markup Language v5

i.i.d. independent and identically distributed

ICI Inter Carrier Interference

IDFT Inverse Discrete time Fourier Transformation

ISI Inter Symbol Interference

LTE Long Term Evolution

MAP Maximum a-posteriori

ML Maximum Likelihood

MLE Maximum Likelihood Estimator

MSE Mean Squared Error

OFDM Orthogonal Frequency Division Multiplexing

PAPR Peak to Average Power Ratio

PC Personal Computer

PDF Probability Density Function

PDP Power Delay Profile

PSD Power Spectral Density

PSK Phase Shift Keying

QAM Quadrature Amplitude Modulation

QPSK Quaternary Phase Shift Keying

SDR Software Defined Radio

SLM Selected Mapping

SNR Signal to Noise Ratio

SSPA Solid State Power Amplifier

TWTA Travelling Wave Tube Amplifier

USRP Universal Software Radio Platform

VDSL Very High Speed Digital Subscriber Line

WAN Wide Area Network

WLAN Wireless Local Area Network

1. Introduction

1.1. Development of Communications Systems

The trend towards transmitting data with ever increasing rates continues both in Wireless Local Area Network (WLAN) as well as celluar Wide Area Network (WAN) applications. Portable smartphones and tablet computers with ever growing computing power and high-definition displays are replacing traditional desktop and mobile Personal Computers (PCs) especially in many home applications. Cloud computing together with modern web standards such as Hypertext Markup Language v5 (HTML5), Cascading Style Sheets v3 (CSS3) and JavaScript now allows for rich web applications that increasingly replace classical desktop applications. Furthermore, the advent of high-definition audio and video streaming services create demand for always-on, low-latency internet connections with ever increasing bandwidth requirements.

One major technology behind this development is multicarrier modulation, especially in form of Orthogonal Frequency Division Multiplexing (OFDM). OFDM is utilized in current and future cellular networks such as Long Term Evolution (LTE) or LTE-advanced, local area networks (802.11a/g/n), digital broadcasting services like Digital Audio Broadcasting (DAB) or Terrestrial Digital Video Broadcasting (DVB-T) and internet access over copper wires like Asymmetric Digital Subscriber Line (ADSL) and more recently also Very High Speed Digital Subscriber Line (VDSL). One reason for todays dominance of OFDM is the computationally cheap implementation of transmitters and receivers by means of the Fast Fourier Transformation (FFT) as well as the relatively simple equalization of frequency selective channels when compared to other multi-carrier or even single carrier methods. Furthermore, OFDM enables fine grained resource allocation in time and frequency which is crucial for high speed cellular standards such as LTE. However, there are some practical issues when implementing OFDM.

1.2. Practical Implementation of OFDM Systems

OFDM can be described as a parallel transmission of multiple orthogonal single carrier signals. Each of the parallel transmissions is refered to as an OFDM subcarrier and modern OFDM systems employ 1024 or more subcarriers. Similar to single carrier systems, each subcarrier has a relatively limited dynamic range. However, due to constructive interference, the sum of all subcarriers exhibits large power spikes that increase the signal dynamic range. Power peaks in OFDM systems can reach 10 dB and more above the average power level. This imposes big issues for the design of power amplifiers in OFDM systems. Development and manufacturing of amplifiers operating linearly over a wide dynamic range is a challenging engineering problem. Furthermore, most amplifiers operate at peak energy efficiency when driven close to saturation. However, in order not to distort power peaks of OFDM signals, an amplifier operating point providing a large power reserve, the so called backoff, has to be chosen. This puts the amplifier into an energy inefficient mode of operation which is especially bad for battery powered devices such as laptops, tablets or smartphones.

In the past, this problem has attracted a lot of attention from the research community and many promising methods have been developed to alleviate nonlinear distortions. Most of these fall in one of three categories. Firstly, predistortion methods introduce a digital nonlinearity in the signal path such that the combined effects of digital and analog nonlinearity yield an overall linear characteristic. Secondly, Peak to Average Power Ratio (PAPR) reduction methods seek to reduce the dynamic range of the OFDM time domain signals. For example, a different data-to-carrier mapping or redundant coding can reduce the dynamic range of the resulting time signal. Finally, it has been shown that even strong nonlinear distortions do not have a large effect on the resulting channel capacity. Based on these results, receiver architectures have been developed that can mitigate the effects of nonlinear distortions to a large degree. There also exist hybrid methods such as clipping and filtering that combine digital predistortion to reduce the dynamic range of the signal and a receiver that can mitigate the resulting signal distortions.

1.3. State of the Art

As stated above, many methods have been developed to cope with the high dynamic range of OFDM signals and the problems that arise when amplifying these signals. These methods usually require knowledge of the nonlinearity characteristic of the amplifier. Special training sequences have been proposed to obtain this knowledge. However, employing such a pilot sequence would reduce the overall data rate. Therefore, it is desireable to obtain knowledge about the nonlinearity characteristic only by observing the data signal at the receiver. This approach is commonly known as blind estimation. To the best knowledge of the author, there exists no comprehensive analysis of blind nonlinearity estimation. The focus and the new contributions of this thesis are in this field.

1.4. Goals of this Work

The following are the three main goals of this work:

1. Show the feasibility of blind nonlinearity estimation by mathematical derivation of estimators and characterization of their performance.
2. Analyze the complexity of the estimators and evaluate their applicability to real-time environments. If required, derive low-complexity algorithms and give an overview of the tradeoffs.
3. Derivations are done using widely accepted nonlinearity models. Show that the resulting algorithms are suitable for use in the real world by testing them with actual hardware amplifiers operated in saturation mode.

1.5. Notation

The following notation is used throughout the thesis.

- h, H: Normal symbols represent scalar values. In cases where a signal could be in time or frequency domain, lower case symbols represent time domain values and upper case symbols represent frequency domain values.
- $\boldsymbol{h}, \boldsymbol{H}$: Bold symbols represent vectors. Upper and lower case rules are similar to scalars.

- $\underline{\boldsymbol{H}}$: Underlined bold symbols represent matrices. They are usually using upper case letters.

- s_k, $[\boldsymbol{s}_\phi]_k$: Both notations represent the k-th element of the vector $\boldsymbol{s}$ resp. $\boldsymbol{s}_\phi$. The first notation is used when the vector itself does not have a subscript as part of its name. It is often used together with a sum or product operator that defines the index variable. If no such operator is present it represents an arbitrary element of that vector, e.g. in cases where an equation is defined per-element. The second notation is used in ambiguous cases, e.g. when the vector itself has a subscript as part of its name.

- $[\underline{\boldsymbol{H}}]_{ij}$: Denotes the element in the i-th row and j-th column of the matrix $\underline{\boldsymbol{H}}$.

1.6. Structure of the Thesis

The thesis is structured as follows:

In **chapter 2**, an introduction to OFDM signals and memoryless nonlinearities is given. The effects of nonlinear distortion of OFDM signals is characterized and an overview of existing mitigation methods is presented.

In **chapter 3**, a blind feedforward nonlinearity estimator based on maximum likelihood estimation is formally derived. The estimation is based solely on the received signal. It is shown that it exhibits prohibitively high computational complexity and low complexity methods are derived for two special cases.

In **chapter 4**, a blind feedback nonlinearity estimator is derived. In addition to the received signal, it also uses the information in the received information bits through remodulation. It is shown that the method offers better estimation performance in a wider range of scenarios with lower computational complexity.

In **chapter 5**, a software defined radio implementation of a feed forward estimator working with signal distorted by real amplifiers is presented.

In **chapter 6**, a summary of the most important results as well as an outlook on future work is given.

2. Nonlinearly Distorted Orthogonal Frequency Division Multiplexing (OFDM) Signals

Multicarrier modulation using orthogonal subcarriers, i.e., OFDM, has become a wide-spread modulation scheme for many mobile radio applications. As such, it forms the basis for all methods derived in this thesis. In this chapter, a short introduction to OFDM, its relevant properties and the effects of nonlinear distortions are given. For a more general introduction to OFDM, there are numerous publications such as [Nee99, BSE04].

In section 2.1, the OFDM modulation scheme is defined and the most relevant statistical properties of the resulting signals are described. Memoryless nonlinearities and their effects on OFDM signals are investigated in section 2.2. Finally, an overview of current approaches to mitigate the negative effects of nonlinearities on OFDM signals is given in section 2.3.

2.1. Introduction to OFDM

2.1.1. From Orthogonality to OFDM

Orthogonality A set of functions ξ_m is called orthogonal if

$$\int_a^b \xi_m(t)\xi_n^*(t)\mathrm{d}t = \begin{cases} 0 & \text{if } m \neq n \\ ||\xi_m||_2^2 & \text{if } m = n\,. \end{cases} \tag{2.1}$$

for appropriate integration boundaries [WJ65]. The set is called orthonormal if $||\xi_m||_2^2 = 1$, which can generally be achieved by means of normalization. Given a weighted sum of orthonormal functions

$$\xi(t) = \sum_{m=0}^{N} d_m \xi_m(t) , \tag{2.2}$$

each weight d_m can be recovered as follows:

$$d_m = \int_a^b \xi(t)\xi_m^*(t)\mathrm{d}t . \tag{2.3}$$

By setting d_m to represent information, this concept is widely used in communications and forms the basis for many modulation schemes and algorithms.

Single-carrier Many typical single-carrier transmission methods employ a Quadrature Amplitude Modulation (QAM) scheme. The time discrete data symbols $d_m \in \mathbb{C}$ are used to weight a train of unit impulses placed equidistantly with the symbol duration T_s. This weighted impulse train is then filtered with a band-limiting pulse shaping filter that has the impulse response $h_p(t)$ to form the baseband signal $s(t)$ as follows:

$$s(t) = \sum_{m=-\infty}^{\infty} d_m \cdot h_p(t - mT_s) . \tag{2.4}$$

If the system is designed to be orthogonal, comparing to (2.2) shows that the set of orthogonal functions is formed from time shifted variants of the pulse shape $h_p(t)$. This is achieved most easily by making the pulses non-overlapping, i.e., $h_p(t) = 0$ if $|t| > T_s/2$. However, there exist certain overlapping pulse shapes that still fulfill the orthogonality criteria, the so-called Nyquist pulses [Nyq28, PS07]. A system in which consecutive symbols do not interfere with one another is called Inter Symbol Interference (ISI)-free.

Multi-carrier Single-carrier systems spread the data symbols over time. Multi-carrier systems add another dimension by spreading out data symbols over time and frequency. A system with N_c subcarriers having center frequencies $f_0, \ldots, f_{N_c-1}$ is given by

$$s(t) = \sum_{m=-\infty}^{\infty} \sum_{l=0}^{N_c-1} S_{l,m} \cdot h_p(t - mT_s) \cdot \mathrm{e}^{j2\pi f_l t} , \tag{2.5}$$

where $S_{l,m}$ denotes the data symbol modulated onto the l-th subcarrier in the m-th time symbol. First it can be seen that a single-carrier system is just a special case

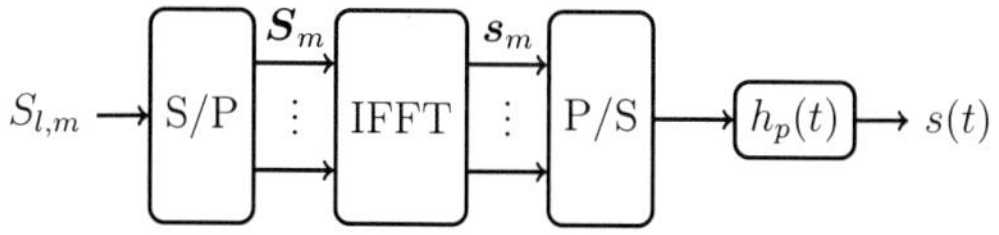

Figure 2.1.: Principal OFDM modulator

with $N_c = 1$ and $f_0 = 0$ Hz. Furthermore, the set of orthogonal functions now consists of the pulse shape $h_p(t)$ shifted in time by integer multiples of the symbol duration and in frequency by $f_0, \ldots, f_{N_c-1}$. If all subcarriers are orthogonal, the system is called Inter Carrier Interference (ICI)-free.

OFDM as an important special case of multi-carrier modulation is described now.

2.1.2. OFDM Modulation and Demodulation

OFDM is a very popular variant of multi-carrier modulation which is usually described and implemented in the digital domain. The simplest form of an OFDM modulator is shown in Figure 2.1. As indicated by the serial/parallel converter, an OFDM symbol $\boldsymbol{S}_m$ consists of N_c data symbols $S_{l;m}, 0 \leq l < N_c$. N_c is called the number of subcarriers of the OFDM symbol since it represents the number of simultaneously transmitted symbols in orthogonal frequency bins. The modulation to different frequencies is done by means of an Inverse Discrete time Fourier Transformation (IDFT). Hence, the time samples $\boldsymbol{s}_m$ belonging to the m-th OFDM symbol are given as:

$$s_m[k] = \frac{1}{\sqrt{N_c}} \sum_{l=0}^{N_c-1} S_{m;l} \cdot \mathrm{e}^{j2\pi \frac{lk}{N_c}} = \sqrt{N_c} \cdot \mathrm{IDFT}\left\{\boldsymbol{S}_m\right\} . \tag{2.6}$$

The factor $\frac{1}{\sqrt{N_c}}$ is a power normalization so that $\mathrm{E}\left\{|s(t)|^2\right\} = \mathrm{E}\left\{|S_{m;l}|^2\right\}$. Since the output of the IDFT consists of N_c time domain samples, the duration T_o of a single OFDM symbol is $T_o = N_c \cdot T_s$. (2.6) shows that the l-th data symbol belonging to each OFDM symbol is digitally mixed to the frequency $\frac{l}{N_c}$. The signal is then converted from the digital to the analog domain by pulse shaping with $h_p(t)$. The digital frequencies $\frac{l}{N_c}$ correspond to $\frac{l}{T_o}$ in the analog domain and hence, the frequency difference between two subcarriers is $\frac{1}{T_o}$ which has been shown [CTL12] to be the minimal required spacing for the subcarriers to be orthogonal. Hence, using the IDFT not only ensures orthogonality in the frequency domain avoiding inter-carrier interference, it also requires the least frequency resources possible to do so. Furthermore, using a Nyquist pulse for $h_p(t)$ ensures orthogonality in the time domain avoiding inter-symbol interference.

Assuming that there are no noise or other disturbances, i.e., the received signal $r_m(t) = s_m(t)$, the received samples

$$r_m[k] = s_m[k]\,. \tag{2.7}$$

OFDM demodulation is done by inverting (2.6), i.e.,

$$R_{m;l} = \frac{1}{\sqrt{N_c}} \sum_{k=0}^{N_c-1} r_m[k] \cdot \mathrm{e}^{-j2\pi\frac{lk}{N_c}}\,. \tag{2.8}$$

(2.8) can be expressed by means of the Discrete time Fourier Transformation (DFT) as follows

$$\boldsymbol{R}_m = \frac{1}{\sqrt{N_c}} \cdot \mathrm{DFT}\left\{\boldsymbol{r}_m\right\}\,, \tag{2.9}$$

where $\boldsymbol{r}_m$ and $\boldsymbol{R}_m$ represent vectorized notations of the received samples $r_m[k]$ and symbols $R_{m;l}$.

It shows, that OFDM modulation and demodulation can be achieved by means of IDFT and DFT. The Fast Fourier Transformation (FFT) and its inverse are very efficient algorithms for implementing these transforms and readily available on many development platforms.

2.1.3. Frequency selective channels

A mobile radio channel can be roughly characterized by two properties. The coherence time T_{coh} is the time in which the impulse response of the channel can be assumed to be constant. The coherence bandwidth B_{coh} is the bandwidth in which the frequency response of the channel can be assumed to be constant so that signals with smaller bandwidth experience only flat fading [Gar07].

One of the main reasons for using OFDM is the small bandwidth f_{sc} of each individual subcarrier. If $f_{sc} < B_{coh}$, each subcarrier is affected by approximately frequency flat fading that allows for relatively simple equalization. If the total bandwidth of the system is fixed to be B, $f_{sc} = \frac{B}{N_c}$ depends on the amount of subcarriers and can be made arbitrarily small by increasing the amount of subcarriers. Care should be taken that $\frac{N_c}{B} = T_o < T_{coh}$ or the fading will become time-selective during one OFDM symbol. There is a tradeoff between small subcarrier bandwidth and symbol duration given B is fixed. This means that there are channels for which $T_o < T_{coh}$ and $f_{sc} < B_{coh}$ cannot both be achieved. However, in many cases this is possible and in the following, fulfillment of both conditions is always assumed.

Since the channel coherence time is assumed to be longer than the duration of one OFDM symbol, the channel is modelled by a time-invariant impulse response $h(\tau)$ for the duration of one symbol. Subsequent symbols might see a different channel realization. For most practical channels the impulse response vanishes after some time, i.e., $h(\tau) = 0$ for some $\tau > \tau_c$. To satisfy $f_{sc} < B_{coh}$, it is required that $\tau_c < T_o$. Sampling the impulse response with the same sampling time $T_s = \frac{T_o}{N_c}$ as the OFDM system yields the vector

$$\boldsymbol{h} = \left[h(0), h(T_o \frac{1}{N_c}), h(T_o \frac{2}{N_c}), \ldots, h(T_o \frac{l_c - 1}{N_c}), 0, \ldots, 0 \right] , \qquad (2.10)$$

with l_c being the length of the impulse response counted in number of samples. Furthermore, let $\boldsymbol{H} = \text{DFT}\{\boldsymbol{h}\}$.

Let $\boldsymbol{S}$ be a transmitted OFDM symbol and $\boldsymbol{H}$ the frequency response of the channel. Then, the resulting signal in frequency domain is

$$\boldsymbol{R} = \boldsymbol{S} \odot \boldsymbol{H} , \qquad (2.11)$$

where $\odot$ denotes element-wise vector multiplication. For brevity, the time index m has been omitted. Each subcarrier of the source symbol is multiplied with just one complex element of the channel's frequency response, yielding frequency flat fading and allowing for very simple equalization. However, (2.11) is based on the convolution theorem of the Fourier Transform and in order to hold for the DFT, the convolution needs to be circular as follows:

$$\boldsymbol{r} = \boldsymbol{s} \circledast \boldsymbol{h} . \qquad (2.12)$$

In a real system, convolution happens in a linear way. Due to channel memory, this causes OFDM symbols to leak energy into subsequent symbols yielding ISI. Furthermore, (2.12) does not hold anymore. In almost all OFDM systems, this problem is solved by introducing the Cyclic Prefix (CP).

Circular and linear convolution become indistinguishable after l_c samples for a channel with length l_c. Adding a cyclic prefix with length $l_{cp} > l_c$ as follows

$$\boldsymbol{s}_{cp} = [s[N_c - l_{cp}], \dots s[N_c - 1], s[0], \dots s[N_c - 1]] \tag{2.13}$$

ensures that the last N_c samples of $\boldsymbol{s}_{cp}$ after it was subject to linear convolution with $\boldsymbol{h}$ are equal to a circular convolution of $\boldsymbol{s}$ and $\boldsymbol{h}$. The cyclic prefix is typically designed to be larger than the maximum expected delay spread of the channel. On the receiver side, the cyclic prefix can be discarded or the additional information it contains can be used to aid in synchronization, frequency offset correction and other tasks [SvdBB95]. Using the cyclic prefix for synchronization works especially well with unique word OFDM where the cyclic prefix is replaced with an a-priori known sequence [HHH10]. In the following, the cyclic prefix is omitted from the derivations. It is implicitely assumed to be present, long enough to avoid ISI and discarded at the receiver. A more in-depth explaination of the cyclic prefix and especially the implications on data rates can be found in [DOB+09].

2.1.4. Signal Properties

For the coming analysis of nonlinearly distorted OFDM systems, the statistical signal properties of the time domain signals are crucial. The definition of an OFDM system (2.6) states that in time domain the signal is a superposition of N_c complex waves, i.e., the subcarriers of the system. Each complex wave is weighted by a complex-valued constant that represents a data symbol being transmitted on one subcarrier of one specific OFDM symbol. Firstly, the two theoretical extreme cases of just one subcarrier and infinitely many subcarriers are investigated.

One subcarrier system An OFDM system with only one subcarrier is similar to a single-carrier system that is shifted to a non-zero center frequency. The amplitude is constant during each symbol and possible amplitudes match the used symbol alphabet. Consequently, the Peak to Average Power Ratio (PAPR) is also defined by the symbol alphabet and can be 0 dB if a constant amplitude Phase Shift Keying (PSK)-type alphabet and rectangular pulse-shaping is used.

Many subcarrier system For $N_c \to \infty$, each time domain sample is formed by superposition of all weighted complex waves. Since each wave is randomly weighted with a symbol from the alphabet, this is equal to summing up N_c independent random variables. Therefore, using the central limit theorem, it follows that the probability distribution of the time domain samples $s[k]$ converges to a complex normal distribution for large N_c, i.e.,

$$s[k] \rightarrow CN(0, \sigma_s) . \tag{2.14}$$

$\sigma_s = \mathrm{E}\{|S[\cdot]|^2\}$ is the average symbol power of the modulation alphabet. If all subcarriers are used and no per-subcarrier power allocation (e.g., water-filling) is performed, then all time-domain samples are independent and identically distributed (i.i.d.). If there is a guardband (i.e., unused subcarriers) or unequal power allocation per subcarrier, the samples belonging to one OFDM symbol become correlated. Details on that are given later when required. Real and imaginary parts of the samples can also be considered independent which yields a Rayleigh distribution for the signal amplitudes and a uniform distribution for the signal phases.

Peak to Average Power Ratio (PAPR) Of special interest for this thesis is the ratio of peak power to average power κ which for a single OFDM symbol is defined as:

$$\kappa = \frac{\max\left(|s[k]|^2\right)}{\sigma_s^2} . \tag{2.15}$$

Specifically this is the PAPR of the time-discrete complex baseband samples $s[k]$. It is known that the PAPR of the continuous-time baseband signal $s(t)$ is usually higher. However, it has been shown [WPP08] that the difference is bounded and hence, the results apply similarly.

The main consequence of normally distributed time domain samples is a very high PAPR. In the case of no guardband, $\boldsymbol{s}$ consists of i.i.d. zero-mean complex normally distibuted samples with variance σ_s^2. Therefore, the amplitudes are i.i.d. Rayleigh distributed, i.e. $|s[k]| \rightarrow Rayleigh\left(\frac{\sigma_s}{\sqrt{2}}\right)$. Since the PAPR depends on the random vector $\boldsymbol{s}$, it is a random variable. The probability that the PAPR is smaller than a certain value κ_0 is given as

$$\begin{aligned} Pr\left(\kappa < \kappa_0\right) &= Pr\left(\forall \frac{|s[k]|^2}{\sigma_s^2} < \kappa_0\right) \\ &= \prod_{k=0}^{N_c-1} Pr\left(\frac{|s[k]|^2}{\sigma_s^2} < \kappa_0\right) \\ &= \left[F_{rayl}\left(\sqrt{\kappa_0}, \sigma - \frac{1}{\sqrt{2}}\right)\right]^{N_c} , \end{aligned} \tag{2.16}$$

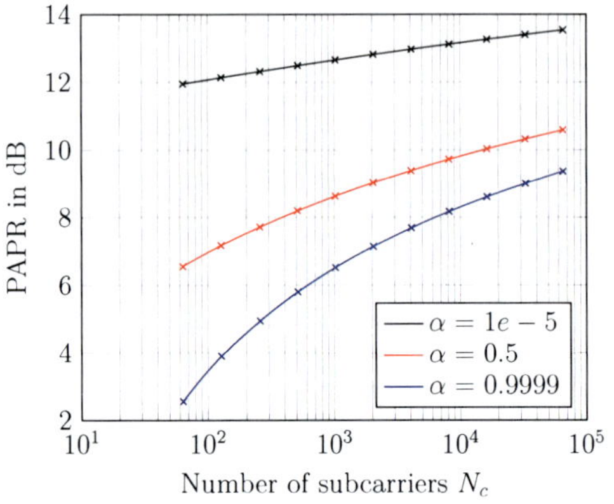

Figure 2.2.: PAPR of OFDM symbols, markers where N_c is power of two

where $F_{rayl}(\cdot)$ represents the Cumulative Distribution Function (CDF) of the Rayleigh distribution which is defined as

$$F_{rayl}(x, \sigma) = 1 - \mathrm{e}^{-\frac{x^2}{2\sigma^2}} . \tag{2.17}$$

Hence, the final result for $Pr(\kappa < \kappa_0)$ can be written as

$$Pr(\kappa < \kappa_0) = \left[1 - \mathrm{e}^{-\kappa_0}\right]^{N_c} \tag{2.18}$$

which has been shown before in [MBFH97]. Figure 2.2 gives an overview of the expected PAPR for different amounts of subcarriers. The three curves show the PAPR that is not exceeded by 0.001%, 50% and 99.99% of all OFDM symbols. Generally, the more subcarriers the system has, the higher the power peaks are. As an example, for Long Term Evolution (LTE) with 1200 active subcarriers, almost all symbols will have a PAPR between 6.5 dB and 13 dB.

As stated earlier, amplifiers need to be designed to exhibit a large linear range to amplify these signals without distortion. The following section describes the consequences when this condition is not met and OFDM signals are nonlinearly distorted.

Validity of Gaussian assumption Since the time-domain behaviour of OFDM signals differs greatly between the extreme cases of one and infinitely many subcarriers, the question about the amplitude distribution for a finite amount of subcarriers $N_c > 1$ arises. This is a well researched topic and the answers differ. In [Zil07], Zillmann gives empirical examples that show significant deviation of the densities of real and imaginary signal components from the normal distribution for a 64 subcarrier OFDM system with Quaternary Phase Shift Keying (QPSK) modulation. For higher-order modulation schemes such as 64-QAM the densities match more closely. Furthermore, for 1024 subcarriers the distributions seem to match very well even for QPSK modulation. In [DTV00], Dardari theoretically analyzed the influence of nonlinearities on OFDM systems. For 64 subcarriers, good matching between simulation and analysis with normally distributed samples was observed in terms of the resulting Bit Error Rate (BER). Recently in [AD12], Araujo investigated the validity of the Gaussian assumption for modelling the effects of nonlinearities on OFDM systems and found that for more than 100 subcarriers, the differences between the exact behaviour and the Gaussian approximation become insignificant. Many other publications on the topic implicitely assume the samples as Gaussian distributed and usually no significant discrepances between theory and simulation arise.

Based on this research the author considers it safe to assume that for $N_c >=$ 128 subcarriers, the Gaussian assumption is reasonably exact for investigating nonlinearities in an OFDM context. Since modern OFDM systems almost always have more subcarriers, this assumption is also realistic in the light of currently used communications systems. For the rest of this work, the Gaussian assumption is always assumed as valid.

2.2. Memoryless Nonlinearities and OFDM signals

It was pointed out in the previous sections that OFDM signals exhibit a large PAPR and why this is problematic for amplifier design. There exist a large amount of models for describing the behavior of amplifiers in the digital baseband that differ mostly by the characterization of frequency selective memory effects. A good overview is given in [PM05]. The relevant model class for this thesis is the memoryless nonlinearity which shall be described now.

2.2.1. Memoryless nonlinearities

Definition For a memoryless nonlinearity, the output at a certain time instant depends only on the input at exactly the same time instant. Furthermore, only the input signal's amplitude is relevant, not the phase. Since the thesis uses models in the equivalent complex baseband, the nonlinearities are specified as baseband models as well. To that end, a memoryless nonlinearity is defined as follows:

$$g(x) = g_A(|x|) \cdot \exp\left(\arg(x) + g_\phi(|x|)\right) . \tag{2.19}$$

The characteristic consists of an Amplitude to Amplitude (AM/AM) distortion $g_A(|x|)$ and an Amplitude to Phase (AM/PM) distortion $g_\phi(|x|)$.

Strict and quasi memoryless nonlinearities There is often some confusion about the exact meaning of memoryless. This arises from the AM/PM distortion that does not show up for strictly memoryless amplifers. Power amplifiers are usually the last element in a transmission chain before emitting the radio waves over the antenna. Hence, the amplifier operates on a passband signal centered around the carrier frequency f_c. Let $\widetilde{s}(t)$ and $\widetilde{r}(t)$ be the passband representations of $s(t)$ and $r(t)$:

$$\widetilde{s}(t) = \Re\left\{s(t) \cdot \exp(j2\pi f_c t)\right\} \tag{2.20}$$

$$r(t) = \frac{1}{\sqrt{2}} \cdot \left(\widetilde{r}(t) + j\mathcal{H}\left\{\widetilde{r}(t)\right\}\right) \cdot \mathrm{e}^{-j2\pi f_c t} \tag{2.21}$$

Here, $\mathcal{H}\{\cdot\}$ represents the Hilbert transform and $\left(\widetilde{r}(t) + j\mathcal{H}\left\{\widetilde{r}(t)\right\}\right)$ is the analytic representation of $\widetilde{r}(t)$ which is complex valued and has the same spectrum as $\widetilde{r}(t)$ for the positive frequencies but all negative frequencies removed. Hence, after down-conversion, a baseband signal around 0 Hz remains. Now, let $\widetilde{g}(\cdot)$ be a memoryless passband nonlinearity that is represented as a power series as follows:

$$\widetilde{r}(t) = \widetilde{g}(\widetilde{s}(t)) = \sum_{l=0}^{\infty} \widetilde{b}_l \widetilde{s}^l(t) . \tag{2.22}$$

Since the amplifier is in the analog/physical domain, all signals and coefficients $\widetilde{b}_l$ have to be real valued. It can then be shown that the baseband equivalent representation is given as [ZQDR05]:

$$r(t) = g(s(t)) = \sum_{l=0}^{\infty} b_{2l+1} |s(t)|^{2l} s(t) \tag{2.23}$$

with

$$b_{2l+1} = \frac{1}{2^{2l}} \binom{2l+1}{l} \tilde{b}_{2l+1} \, . \tag{2.24}$$

It shows that for a memoryless passband nonlinearity, the resulting power series representation of the baseband equivalent nonlinearity has only real coefficients, too. That means, for such a nonlinearity there is only AM/AM but no AM/PM distortion. Then, what real phenomenon does an AM/PM nonlinearity represent?

The answer is very similar to transferring a narrow-band signal over a frequency selective channel. As stated in section 2.1.3, when the signal bandwidth B is smaller than the channel coherence bandwidth B_{coh}, the fading becomes flat and can be represented as a multiplication with a single complex coefficient. Similarly, if the nonlinearity has a memory, but it's length τ_g is small compared to variations in the complex envelope of the signal, i.e., $s(t+\tau_g) \approx s(t)$, the resulting instantaneous effect on the signal is similar to frequency flat fading and can be represented as a single complex multiplication at each time instant. This is exactly what a combination of AM/AM and AM/PM nonlinearities does. A formal and detailed derivation of this is given in [RZ02].

In summary, a strictly memoryless nonlinearity will only exhibit AM/AM characteristics. However, if the amplifier exhibits a very short memory, it can be represented as memoryless with AM/AM and AM/PM distortions. Hence, this class of amplifiers is called quasi-memoryless in the literature.

2.2.2. Common models for memoryless nonlinearities

In the following, three very common models of memoryless nonlinearities are presented: The Solid State Power Amplifier (SSPA) model, the Travelling Wave Tube Amplifier (TWTA) model and the soft limiter.

SSPA Model The SSPA model was proposed in 1991 by Rapp [Rap91] as a model that characterizes the properties of solid state amplifiers better than the other models at that time. These properties are an improved linearity in the small signal region, the tendency of the transfer function to a maximum output amplitude for large

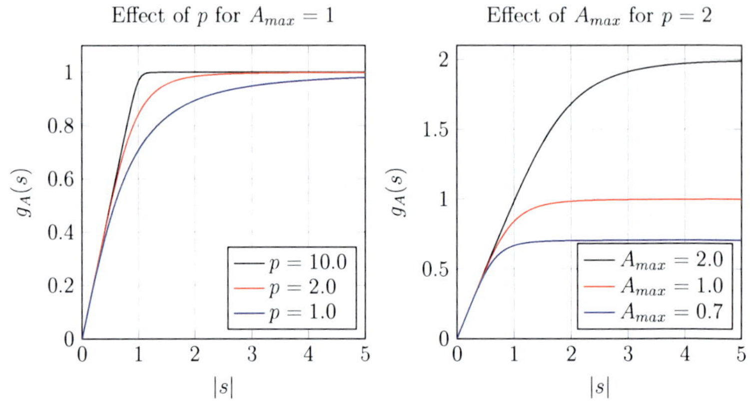

Figure 2.3.: AM/AM characteristic of SSPA model

inputs and the mostly negligible AM/PM distortion of solid state amplifiers. To that end, the model is defined as:

$$
\begin{aligned}
g_A(s, A_{max}, p) &= \frac{|s|}{\left[1 + \left(\frac{|s|}{A_{max}}\right)^{2p}\right]^{\frac{1}{2p}}} \\
g_\phi(s) &= 0\,.
\end{aligned}
\tag{2.25}
$$

Figure 2.3 shows the resulting characteristic for different parameter combinations. The maximum output amplitude of the amplifier in saturation is A_{max} and p defines how smooth the transition between the linear and saturated part of the characteristic is. For small s, the model is linear and for large inputs it saturates towards A_{max}. The model does not exhibit AM/PM distortion to emphasize the negligible phase distortions of solid state amplifiers. It is therefore a strictly memoryless model.

TWTA Model The TWTA model was proposed in 1981 by Saleh [Sal81] to characterize the properties of travelling wave tube amplifiers. Today it is mainly used to model amplifiers in satellite communications. These amplifiers exhibit a strong AM/PM distortion and their amplitude characteristic has a maximum and tends to

	AM/AM	AM/PM
$\|s\| \to 0$	linear behavior	quadratic behavior
$\|s\| \to \infty$	$g_A(s) \to 0$	$g_\phi(s) \to \frac{\alpha_\phi}{\beta_\phi}$
Maxima	$\|s_0\| = \frac{1}{\sqrt{\beta_A}}$, $g_A(s_0) = \frac{\alpha_A}{2\sqrt{\beta_A}}$	None

Table 2.1.: Extreme cases of TWTA nonlinearity

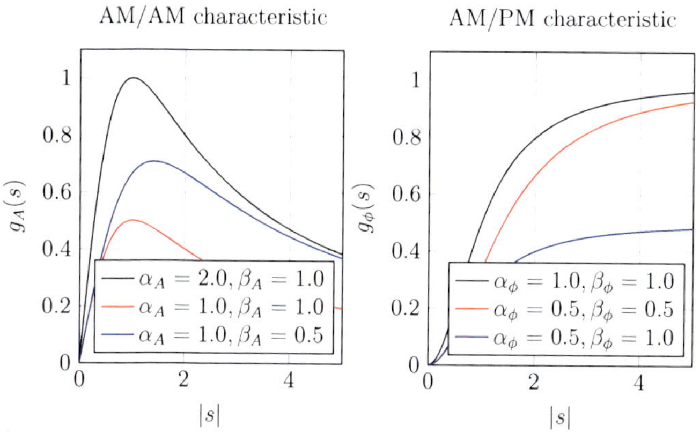

Figure 2.4.: Characteristic of TWTA model

decrease for larger inputs. The model is defined as follows:

$$g_A(s, \alpha_A, \beta_A) = \frac{\alpha_A |s|}{1 + \beta_A |s|^2} \tag{2.26}$$

$$g_\phi(s, \alpha_\phi, \beta_\phi) = \frac{\alpha_\phi |s|^2}{1 + \beta_\phi |s|^2}\,. \tag{2.27}$$

Since the relationship between the parameters and the shape of the resulting curves is not obvious, table 2.1 shows the behaviour for some special cases and Figure 2.4 shows exemplary AM/AM and AM/PM characteristics. The parameter α_A represents the small system gain of the model. In this thesis, normalized gain is always assumed without loss of generality, hence from now on $\alpha_A = 1$.

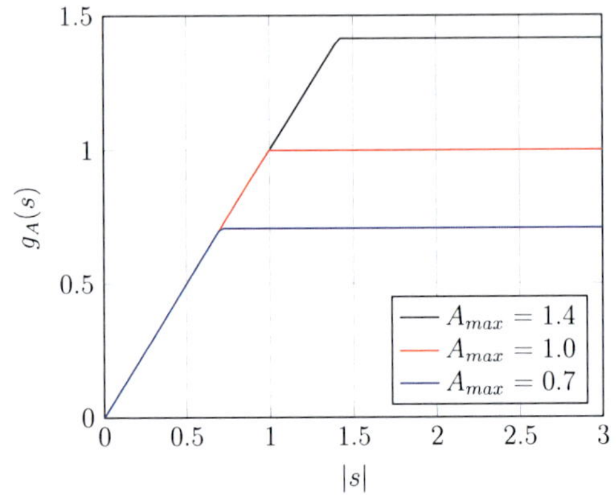

Figure 2.5.: Characteristic of soft limiter model

Soft Limiter The soft limiter is an idealized model for an amplifier with clipping. Its characteristic is perfectly linear until saturation A_{max} is reached. It is defined as:

$$g_A(s, A_{max}) = \begin{cases} |s| & \text{if } |s| \leq A_{max} \\ A_{max} & \text{if } |s| > A_{max} \end{cases} \tag{2.28}$$

$$g_\phi(s) = 0 \tag{2.29}$$

Figure 2.5 shows several exemplary characteristics. The model is an extreme case of the SSPA model for $p \to \infty$.

2.2.3. Nonlinearly distorted OFDM signals

In the last sections OFDM signals and memoryless nonlinearities have been introduced. In this section it is investigated how an OFDM signal is affected by a memoryless nonlinearity. Most of the following results are not limited to OFDM signals but apply in general to bandlimited normally distributed processes. As before, let $s(t)$ be the time domain representation of an OFDM symbol, $g(s)$ a memoryless nonlinearity and $r(t) = g(s(t))$ the signal after nonlinear distortion.

In-band interference and out-of-band leakage From the power series definition (2.23) follows that memoryless distortions have a linear multiplicative and a nonlinear

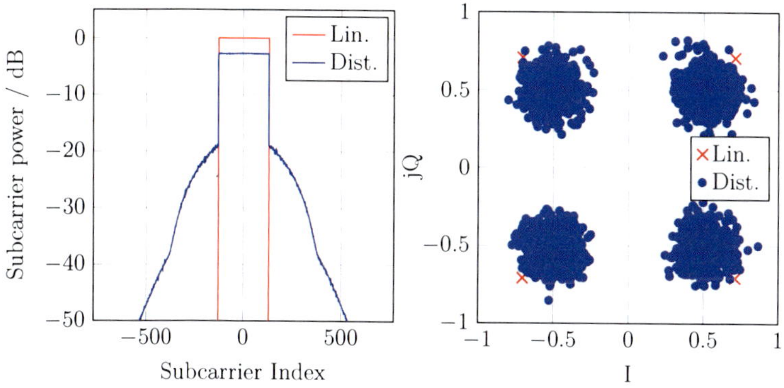

Figure 2.6.: Out-of-band radiation and in-band interference of a 256 subcarrier OFDM system distorted by an SSPA nonlinearity with $p = 2$ and $A_{max} = 0$ dB, QPSK modulation on each subcarrier

additive component. This becomes more visible by rewriting the equation as:

$$r(t) = g(s(t)) = b_1 s(t) + \underbrace{s(t) \cdot \sum_{l=1}^{\infty} b_{2l+1} |s(t)|^{2l}}_{s_g(t)}, \tag{2.30}$$

where the coefficients $b_l \in \mathbb{C}$ to encompass AM/PM distortions as well. Let $R(f)$, $S(f)$ and $S_g(f)$ be the Fourier Transforms of $r(t)$, $s(t)$ and $s_g(t)$. Then, the resulting spectrum is given as:

$$R(f) = b_1 S(f) + S(f) * S_g(f). \tag{2.31}$$

Next to the scaled original spectrum, the output spectrum also contains the convolution of $S(f)$ and $S_g(f)$. Using the properties of the convolution reveals that the bandwidth of $R(f)$ is the sum of the bandwidths of $S(f)$ and $S_g(f)$.

Two effects can be observed. The total bandwidth of the signal is increased resulting in out-of-band leakage. If this happens on the transmitter side, the leakage may lead to violations of the spectral masks which limit the allowed emissions into neighboring channels. Furthermore, the in-band parts of the additive distortion term leads to ICI and therefore decreased Signal to Noise Ratio (SNR). Figure 2.6 shows the Power Spectral Density (PSD) and resulting constellation diagram of a 256 subcarrier OFDM system using QPSK to modulate each carrier that is distorted by a SSPA nonlinearity. To capture the out-of-band effects, the system was simulated with 32-fold oversampling to get a good approximation of the actual continuous time equivalent baseband signal passing through the nonlinearity model. As predicted, there is significant leakage in the adjacent frequency bands as well as a reduction in average energy. Furthermore, the constellation plot shows that the nonlinearity induced a large amount of interference.

While out-of-band leakage is an immensely important problem especially in mobile cellular communications, there are areas in wired and wireless communications where it is less of an issue. For example, the Deutsche Forschungs Gesellschaft (DFG) Highly Adaptive Energy-Efficient Computing (HAEC) project [FNL12] investigates wireless board-to-board communications in the 100-300 GHz band. Due to highly directed beams and very short range, interference is less of an issue and out-of-band distortion causes less disturbances in other links. Similarly, in the 3D Chip Stack Interconnects (3DCSI) project [FuHLF13], interconnects on a chip and between chips on a board are investigated where, again, out-of-band radiation is less of an issue while amplifier efficiency is crucial. The interested reader can find an in-depth analysis of out-of-band radiation in OFDM systems including an analytic derivation of the distorted signal's PSD in [CMP99,CP02]. For the in-band interference however, there is a very important theorem.

The Bussgang decomposition In 1952, Bussgang presented a very useful theorem [Bus52]. It states that if a zero-mean normally distributed process $s(t)$ is input into a memoryless nonlinearity $g(s)$, the cross-correlation $R_{sr}(\tau)$ of $s(t)$ with the output $r(t) = g(s(t))$ is proportional to the autocorrelation $R_{ss}(\tau)$ of the input process:

$$R_{sr}(\tau) = \alpha R_{ss}(\tau) . \tag{2.32}$$

In the original publication, the theorem was proven for time-continuous real-valued stationary normally distributed processes undergoing only AM/AM distortions. In that case, $\alpha = \mathrm{E}\{g'(s(t))\}$. The application to time-discrete processes is straight forward. Over time, different authors extended the scope of the theorem to more general applications. A major contribution was done by Minkoff in 1985. In his publication [Min85] he extended the scope not only to complex-valued normally distributed processes, but to any zero-mean complex valued process that exhibits circular symmetry, i.e., the value of the Probability Density Function (PDF) only depends on the amplitude but not the phase of the argument. Furthermore, Minkoff extended the theorem to encompass AM/PM nonlinearities and showed that it is impossible for an AM/PM nonlinearity to improve the post nonlinearity SNR. It has therefore been shown impossible that careful design of the AM/PM nonlinearity can ease the negative effects of an AM/AM nonlinearity.

It can be seen from (2.32) that by setting $\tau = 0$, α is given as:

$$\alpha = \frac{\mathrm{E}\{\boldsymbol{s}\boldsymbol{r}^*\}}{\mathrm{E}\{\boldsymbol{s}\boldsymbol{s}^*\}} = \frac{\mathrm{E}\{\boldsymbol{s}g^*(\boldsymbol{s})\}}{\sigma_s^2} \tag{2.33}$$

For nonlinearities that exhibit only AM/AM distortions, α will be real-valued whereas for AM/PM distortions, α will be complex valued. Equation (2.32) allows a stochastic description of the output process after the nonlinearity as follows:

$$\boldsymbol{r} = \alpha \boldsymbol{s} + \boldsymbol{n_d}\,. \tag{2.34}$$

The input signal is scaled with α and superimposed by a noise-like signal $\boldsymbol{n_d}$ that is uncorrelated to $\boldsymbol{s}$. $\boldsymbol{n_d}$ is usually not normally distributed and finding the distribution is very difficult in most cases. However, since both $\boldsymbol{r}$ and $\boldsymbol{s}$ are zero-mean, the same holds for $\boldsymbol{n_d}$. The average power σ_r^2 of the output process $\boldsymbol{r}$ is given as:

$$\sigma_r^2 = \mathrm{E}\{\boldsymbol{r}\boldsymbol{r}^*\} = \mathrm{E}\{g(\boldsymbol{s})g^*(\boldsymbol{s})\} = \mathrm{E}\left\{|g(\boldsymbol{s})|^2\right\}\,. \tag{2.35}$$

Assuming a normalized nonlinearity, it is a valid assumption that $|g(s)| \leq |s|$. Under this condition $\sigma_r^2 \leq \sigma_s^2$ so that the nonlinearity reduces the average output power. From (2.34) it also follows that the average output power is given by:

$$\sigma_r^2 = \mathrm{E}\{\boldsymbol{r}\boldsymbol{r}^*\} = \mathrm{E}\{(\alpha\boldsymbol{s} + \boldsymbol{n_d})(\alpha^*\boldsymbol{s}^* + \boldsymbol{n_d}^*)\} \tag{2.36}$$

$$= \mathrm{E}\left\{|\alpha\boldsymbol{s}|^2\right\} + \underbrace{\mathrm{E}\{\alpha\boldsymbol{s}\boldsymbol{n_d}^* + \alpha^*\boldsymbol{s}^*\boldsymbol{n_d}\}}_{0} + \mathrm{E}\left\{|\boldsymbol{n_d}|^2\right\} \tag{2.37}$$

$$\sigma_r^2 = |\alpha|^2\sigma_s^2 + \sigma_{nd}^2\,. \tag{2.38}$$

Combining both equations, the power of the additive distortion term $\boldsymbol{n_d}$ is given as:

$$\sigma_{nd}^2 = \sigma_r^2 - |\alpha|^2\sigma_s^2 = \mathrm{E}\left\{|g(\boldsymbol{s})|^2\right\} - |\alpha|^2\sigma_s^2\,. \tag{2.39}$$

Using this, the signal quality degradation of the nonlinearity alone can be characterized by the post-nonlinearity SNR ρ_{nd} as follows:

$$\rho_{nd} = \frac{|\alpha|^2 \sigma_s^2}{\sigma_{nd}^2}. \tag{2.40}$$

Comparing (2.34) with Figure 2.6 shows that the observations match the predictions again. Following a similar line of thought as in section 2.1.4, the frequency domain representation $\boldsymbol{N_d}$ of $\boldsymbol{n_d}$ is modeled as zero-mean normally distributed due to the central limit theorem.

While (2.34) looks very similar to (2.30), the important difference is that there is no correlation and hence no linear dependency between $\boldsymbol{n_d}$ and $\boldsymbol{s}$ which is not applying to (2.30). For that reason, $\alpha \neq b_1$ except for some specially crafted cases. Furthermore it should be noted that $\boldsymbol{n_d}$ is not a truly random variable. Instead, it is fully dependent on $\boldsymbol{s}$ and $g(s)$ and can be reproduced if both are known which will be useful for nonlinearity mitigation as discussed later.

2.2.4. Conclusions

In this section, the influence of memoryless nonlinear distortions on OFDM signals have been investigated. The main effects are:

- Out-of-band power leakage
- In-band distortions (ICI)
- Reduction of average output power

Missing from this is a derivation of how the PDF of the input signal changes after the nonlinearity. This will be investigated in chapter 3 since it is very important for the blind estimation methods. However, before jumping to that, a short overview of concepts for the mitigation of nonlinear distortions is given.

2.3. Concepts for nonlinearity mitigation

Over time, the high PAPR of OFDM signals and the resulting problems with nonlinear amplifiers have attracted a lot of attention from the research community. Many different algorithms have been developed to mitigate this problem and most of them can be categorized into one of the following classes:

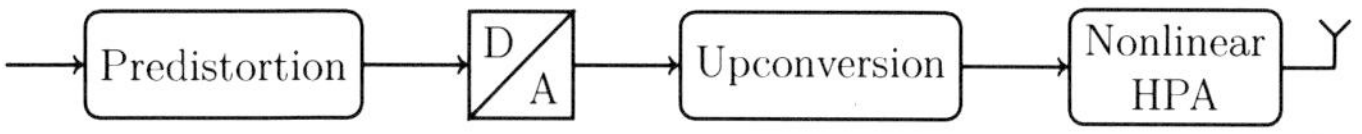

Figure 2.7.: Structure of a transmitter with digital predistortion

- Amplifier linearization through predistortion
- PAPR reduction
- Nonlinearity aware receivers

In the following, the principle of each class is presented.

2.3.1. Linearization through predistortion

The idea of predistortion is to use a preceeding nonlinearity to linearize the power amplifier's characteristic. By careful design it is possible that both nonlinearities combined inflict much less nonlinear distortions on the signals and hence the resulting characteristic is closer to a linear one. Predistortion can be implemented in the digital [D'A01] or analog [YYP00, SKH12] domain but digital predistortion is commonly agreed as more cost-effective since it does not require additional components in the analog frontend. Its basic structure is shown in Figure 2.7.

To design the predistorter, the characteristic of the power amplifier needs to be known. This is problematic since it is an analog component and subject to effects such at aging, production tolerances or circuit temperature which slightly change the electrical properties and therefore the characteristic. Using a static model for the nonlinear power amplifier can result in a residual nonlinearity. For that reason, some authors propose an analog calibration circuitry to account for these effects [BS11].

Figure 2.8 shows an example for the predistortion of a SSPA nonlinearity by means of characteristic inversion. The predistorter is able to extend the linear range of the amplifier significantly. However, the method has a downside. The maximum output level of the actual amplifier can never be surpassed. Therefore, predistortion alone cannot fully mitigate the nonlinear characteristic of amplifiers driven deep into saturation. Instead, residual nonlinearities will remain for very high power peaks. Furthermore, when linearizing a typical characteristic going from a linear to a saturation range, predistorters have the tendency to increase the dynamic range of the signal. Care must be taken not to drive other elements such as D/A-converters

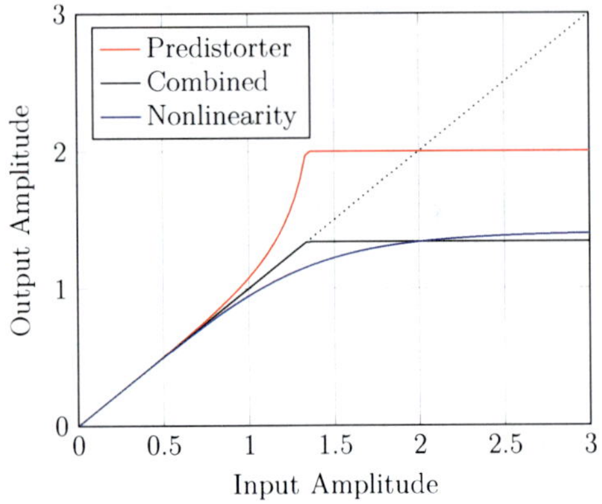

Figure 2.8.: Linearization of a SSPA nonlinearity with $A_{max} = \sqrt{2}$ and $p = 2$, maximum output amplitude of predistorter limited to 2

or mixers into saturation. In Figure 2.8 this is depicted by limiting the maximum output level of the predistorter.

In most cases, predistortion is used to linearize the high-power amplifier in a transmitter. Its big advantage is that it does neither require any changes on the receiver side nor does it generate out-of-band radiation if carefully designed. For these reasons it is a very popular technique for use in cellular base stations where it can significantly improve the power amplifier efficiency [KDJ05].

2.3.2. PAPR reduction

The idea of predistortion is to allow signals with higher PAPR by increasing the effective linear range of the nonlinearity. PAPR reduction techniques approach the problem by reducing the dynamic range of the OFDM signal.

Carrier Rearrangement Recalling Figure 2.2, the PAPR of each individual OFDM symbol is not constant but a random variable and almost all realizations fall into a band of roughly 4-8 dB between highest and lowest PAPR. The reason for the power peaks is the sum of random data weighted complex waveforms that can interfere constructively. Therefore, by changing the random weights it is possible to generate an alternative OFDM symbol with lower PAPR but still carrying the same information.

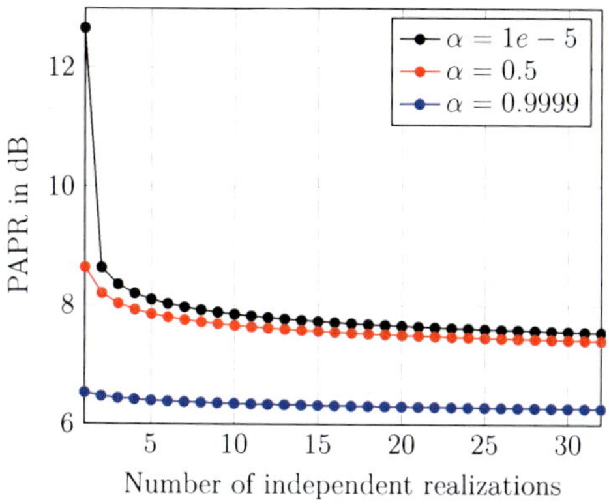

Figure 2.9.: Minimum PAPR of N OFDM symbols, $N_c = 1024$

For example, in the Partial Transmit Sequence method, the symbol is subdivided into carrier groups and each group is subject to a phase rotation which is then optimized to find the symbol with lowest peak power [MH97], reducing the peak by up to 4 dB. A similar method is the Selective Mapping method where N vectors, each containing a random phase rotation are generated and multiplied carrier-wise with the OFDM symbol. Again, the symbol with lowest peak power is transmitted and the method reaches a PAPR reduction of 3-4 dB [BFH96]. Another method employs interleaving the data symbols or data bits to generate a set of OFDM symbols [JT00].

These methods have in common that they do not introduce distortions to the signal. A set of N alternative OFDM symbols is usually generated and the one with the lowest PAPR is transmitted. Assuming that the time domain representations of all generated OFDM symbols can be considered to be independent, Figure 2.9 shows the distribution of the PAPR for different amounts of alternatives. It can be seen that the highest power peaks are reduced by over 4 dB for a very small set of symbols and further reduction requires much larger symbol sets. This matches the cited results that usually reach no more than 3-4 dB PAPR reduction.

The downside of this method is that both transmitter and receiver have to be aware of it. The information about which element of the possible set of symbols has been chosen needs to be transmitted, introducing overhead and reducing the information rate of the system. This transmission needs to happen with especially good error

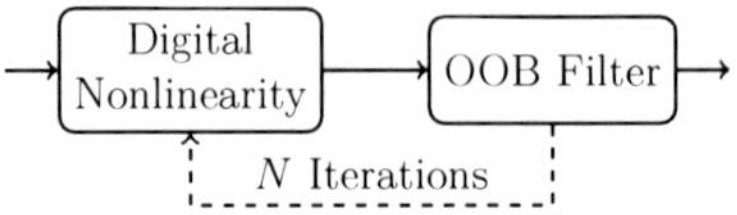

Figure 2.10.: Principle of iterative clipping and filtering

prevention measures since erroneous reception leads to the corruption of a whole OFDM symbol. For some methods, the receiver can estimate the transmitted variant at the expense of higher computational load. Finally, the Selected Mapping (SLM) method requires only very little redundant information which does not even to be specially protected [BMWH01].

Predistortion and PAPR reduction techniques can be combined for an even more efficient usage of the power amplifier [Ryu11].

Iterative Clipping and Filtering Rearrangement techniques reduce the PAPR of OFDM signals without introducing any distortions so that the receiver is not impaired as long as it knows how the rearrangement was done. However, as seen from Figure 2.9, the possible gains are limited and higher gains come at greater expense. The idea of clipping and filtering [Arm02] is to nonlinearly distort the signal but avoid the out-of-band radiation that occurs if an actual power amplifier is the cause of distortion.

The basic principle is shown in Figure 2.10. Firstly, the signal is distorted in the digital baseband. Contrary to digital predistortion, the goal is not to linearize another nonlinearity. Instead, signal peaks are cut off similar to a real amplifier driven into saturation. As shown in Figure 2.6, this induces out-of-band radiation. Therefore, a filter is used to remove all energy from the sidebands. In OFDM systems this is usually done by transforming the signal to the frequency domain and setting the guardband carriers to zero. The filtering will usually increase the PAPR of the resulting signal, diminishing the PAPR reduction. The process is then repeated iteratively until the resulting PAPR has converged to an acceptable level. Furthermore, optimized variants of the algorithm have been presented that reduce the required amount of iterations by means of convex optimization [WL11]. Also, there are approaches to combine this method with rearrangement techniques, see for example [YW06].

The advantages of this method are obvious. The dynamic range of the signal can be limited to a certain desired value and out-of-band radiation is avoided completely.

The downside of this method is that it generates a lot of in-band interference. Therefore, the receiver needs to be aware of the clipping and filtering and the algorithms required to successfully receive a signal which was clipped in this way are usually much more complex than algorithms for unscrambling rearranged OFDM symbols.

2.3.3. Nonlinearity aware receivers

Motivation As stated earlier, according to the Bussgang decomposition, a nonlinearly distorted OFDM time domain signal can be written as follows

$$\boldsymbol{r} = \alpha \boldsymbol{s} + \boldsymbol{n}_d + \boldsymbol{n}\,, \tag{2.41}$$

where $\boldsymbol{n}$ represents Additive White Gaussian Noise (AWGN) with average power σ_n^2 as present in any real radio transmission. The SNR ρ of this signal is given by

$$\rho = \frac{|\alpha|^2 \sigma_s^2}{\sigma_{nd}^2 + \sigma_n^2}\,. \tag{2.42}$$

It can be seen that even for $\sigma_n^2 \to 0$, the resulting ρ will not approach infinity. The result is an error floor. However, this is only true, when an OFDM receiver is used that is not aware of the presence of a nonlinearity and its major noise contribution. As stated earlier, $\boldsymbol{n_d}$ is not actually random, but fully deterministic if $\boldsymbol{s}$ and the nonlinearity $g(s)$ are known. This implies that a specially designed receiver should be able to achieve better performance.

This statement is supported in [ZF05]. In this paper, Zillmann presented a capacity analysis of nonlinearly distorted transmissions. Based on an evaluation of the mutual information between input and output of a memoryless nonlinearity, he showed that the theoretical performance loss is much smaller and that there is no error floor as compared to the case where $\boldsymbol{n_d}$ is assumed as noise.

In his thesis [Zil07], Zillmann provides a comprehensive overview of receiver techniques for nonlinearly distorted OFDM signals ranging from memoryless equalizers, Maximum Likelihood (ML)- and Maximum a-posteriori (MAP)-optimal approaches to iterative receivers based on the Bussgang decomposition. It turned out that for systems with more than $N_c > 64$ subcarriers, the Bussgang-based methods are preferred because they offer low complexity and comparable performance to the ML approaches. For that reason, only this class of receivers is analyzed in this thesis.

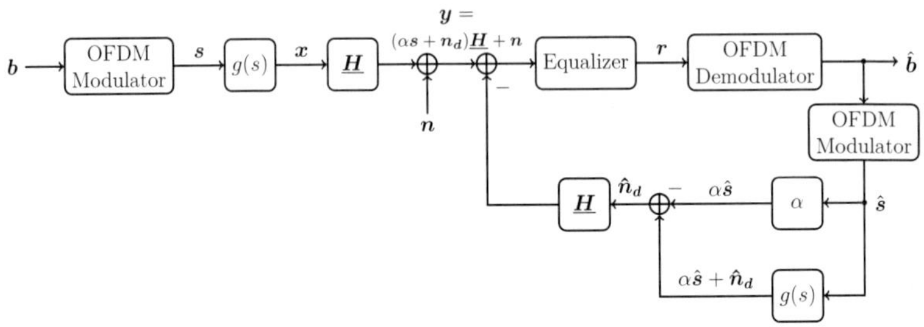

Figure 2.11.: OFDM transmission with Bussgang based decision feedback receiver

Decision Feedback Receiver The idea of a Decision Feedback Receiver (DFR) for nonlinearly distorted OFDM signals has been proposed and analyzed in [THC03]. It is based on the Bussgang decomposition which states that an OFDM time signal $\boldsymbol{s}$ with sufficiently many subcarriers that is subject to memoryless nonlinear distortion $g(s)$ can be decomposed as follows (see (2.34)):

$$\boldsymbol{x} = \alpha \boldsymbol{s} + \boldsymbol{n_d} \,. \tag{2.43}$$

The goal of the DFR is to remove the additive term $\boldsymbol{n_d}$ in order to improve the SNR. While often modeled as noise, $\boldsymbol{n_d}$ is fully deterministic and can be reproduced given that $\boldsymbol{s}$ and the nonlinearity characteristic $g(s)$ are known. Gaining knowledge of the nonlinearity characteristic is investigated in the next chapters and for now, $g(s)$ is assumed as perfectly known.

The basic principle is shown in Figure 2.11. A vector $\boldsymbol{b}$ of source data bits (assumed to be i.i.d. with equal probability of ones and zeros) is modulated using OFDM resulting in the time domain vector $\boldsymbol{s}$. This signal is then nonlinearly distorted, subject to frequency selective fading, modeled by multiplication with the circular convolution matrix $\underline{\boldsymbol{H}}$, and superimposed by AWGN $\boldsymbol{n}$. The signal at the input of the receiver is then given by

$$\boldsymbol{y} = g(\boldsymbol{s})\underline{\boldsymbol{H}} + \boldsymbol{n} = \alpha \boldsymbol{s}\underline{\boldsymbol{H}} + \boldsymbol{n}_d\underline{\boldsymbol{H}} + \boldsymbol{n} \tag{2.44}$$

and in frequency domain by

$$\boldsymbol{Y} = \alpha \boldsymbol{S} \odot \boldsymbol{H} + \boldsymbol{n_d} \odot \boldsymbol{H} + \boldsymbol{N} \,. \tag{2.45}$$

This signal is then equalized, demodulated and detected using standard OFDM techniques. The type of equalizer is not fixed for this receiver but it should be made sure that the complex factor α caused by the nonlinearity is equalized along with the channel. If the channel frequency response $\boldsymbol{H}$ is reconstructed from pilots in the OFDM symbol, this happens implicitely. If the channel is estimated by means of a training sequence that is not subject to nonlinear distortion, this has to be done separately. The result of detection is a vector $\hat{\boldsymbol{b}}$ that is an estimate on the transmitted bits $\boldsymbol{b}$ and may contain errors. These bits are again modulated with exactly the same parameters of the transmitter's OFDM modulator. The resulting $\hat{\boldsymbol{s}}$ is an estimate on the transmit signal $\boldsymbol{s}$ but again might be subject to errors. Using the knowledge of the nonlinearity characteristic $g(s)$ and α, an estimate of the additive distortion $\hat{\boldsymbol{n}}_d$ is calculated:

$$\hat{\boldsymbol{n}}_d = g(\hat{\boldsymbol{x}}) - \alpha\hat{\boldsymbol{x}} = \alpha\hat{\boldsymbol{x}} + \hat{\boldsymbol{n}}_d - \alpha\hat{\boldsymbol{x}} \tag{2.46}$$

After multiplication with the channel matrix $\underline{\boldsymbol{H}}$ the result is subtracted from the input signal of the receiver. In the next iteration, the OFDM receiver will operate on the signal

$$\boldsymbol{y} = \alpha\boldsymbol{s}\underline{\boldsymbol{H}} + (\boldsymbol{n}_d - \hat{\boldsymbol{n}}_d)\underline{\boldsymbol{H}} + \boldsymbol{n} \tag{2.47}$$

and performance will improve if the following inequality holds:

$$\mathrm{E}\left\{|\boldsymbol{n}_d - \hat{\boldsymbol{n}}_d|^2\right\} < \mathrm{E}\left\{|\boldsymbol{n}_d|^2\right\}. \tag{2.48}$$

This process lends itself to iterative application since reducing the power of the additive distortion yields a better estimate of the bits $\hat{\boldsymbol{b}}$. This in turn will result in a better estimation of $\hat{\boldsymbol{n}}_d$.

One important question is when (2.48) holds. When the amount of errors becomes too large, the estimate of the additive distortion might be so far off, that subtracting it introduces additional noise into the system. To the best knowledge of the author, an exact analytical answer to this problem is difficult and has not yet been given. Empirical results show that for common amplifier models presented earlier, (2.48) holds even if the backoff is reduced below 0dB which is much worse than most practical scenarios. In low SNR scenarios, the algorithm does not always converge to the best possible result due to non-recoverable errors that are caused primarily by AWGN contributions.

For the optimal case $\hat{\boldsymbol{b}} = \boldsymbol{b}$, the algorithm fully removes the additive distortion $\boldsymbol{n}_d$. The situation is then equal to linear transmission with an SNR loss of $10 \cdot \log_{10}(|\alpha|^2)$ dB. The DFR treats $\boldsymbol{n}_d$ completely as noise. This is not true since $\boldsymbol{n}_d$ depends fully on the transmitted signal $\boldsymbol{s}$ and the nonlinearity $g(s)$ and therefore contains information about $\boldsymbol{s}$ that this algorithm discards. There are algorithms [RZF06] that employ this information for higher gains. However, those algorithms usually suffer from a much higher computational complexity.

Receivers that are aware of nonlinear distortions have a broad spectrum of usage. In systems with predistortion, they can be used to remove residual nonlinearities. When clipping and filtering is employed or the signal is distorted by an actual nonlinearity in the transmitter, these receivers are required as well. And finally, if the nonlinearity is located in the receiver, it is the only way to remove the signal impairments it caused. Compared to predistortion and rearrangement techniques, the main implementation and computational complexity is on the receiver side. Therefore, these receivers can be used when the transmitter is too weak to implement complex PAPR reduction techniques. However, if the transmitter is simply distorting the signal, out-of-band radiation will occur and care must be taken to not violate spectral masks.

The decision feedback receiver is used throughout the rest of this thesis to benchmark the performance of the estimation algorithms which will provide the DFR with the knowledge about the nonlinearity characteristic $g(s)$.

2.4. Conclusions

In this chapter, a short overview of OFDM transmission and reception methods has been given. The statistical properties of OFDM time domain signals have been analysed especially in terms of their PAPR. Furthermore, an overview of the general properties of memoryless nonlinearities has been given and some common models have been presented. It was shown that OFDM signals that are subject to nonlinear distortion suffer from in-band noise and out-of-band radiation. The resulting signals can be decomposed into a scaled variant of the original signal and an uncorrelated additive noise term by means of the Bussgang decomposition.

In the second part of the chapter, an overview of techniques to deal with the high PAPR of OFDM signals has been given. On the transmitter side, those can be separated into predistortion techniques that try to linearize the nonlinearity characteristic and PAPR reduction techniques. The latter can be divided into rearrangement techniques that try to find alternative signal representations with lower PAPR without distorting the signal and clipping and filtering where the signal is nonlinearly distorted on purpose and the out-of-band radiation is removed by filtering. Finally, a nonlinearity aware receiver has been reviewed that is based on the Bussgang decomposition and removes the additive noise term of the nonlinearity.

With the exception of rearrangement techniques, all methods require knowledge of the nonlinearity. Up to now, this was usually assumed as perfectly known which is not true in most realistic cases. The following chapters will deal with novel methods of gaining knowledge about the nonlinearity characteristic.

3. Blind Feed Forward Estimation

3.1. Introduction

In order to mitigate nonlinear distortions, receiver algorithms require knowledge about the nonlinearity characteristic. When presenting these algorithms, this knowledge is often assumed to be perfect [RZF06, THC03, BZF08]. A practical approach is the transmission of a special preamble with power peaks and average power matching that of the Orthogonal Frequency Division Multiplexing (OFDM) signal [DS07, SDMF07, SSB09]. While this approach is viable, it increases the size of the preamble, effectively lowering the maximum achievable data rate. Furthermore, most mobile standards do not employ a preamble that lends itself to nonlinearity estimation. This excludes receivers for current standards that tolerate nonlinear elements in their analog frontend. These receivers need to derive the required information without any additional pilot signals. This approach is commonly known as blind estimation. To the best of the author's knowledge, not much work has been done in the OFDM context. There is one recent submission that blindly estimates AM/AM nonlinearities using a Kalman filter approach [WUB13].

In this chapter, the problem of blind estimation of amplifier nonlinearities in a feed forward manner is analyzed and novel estimation schemes are proposed. In section 3.2 the system model is presented that forms the basis for the estimation. In section 3.3 the general estimation problem is formulated, the maximum likelihood estimator is derived and problems with practical implementations are discussed. In sections 3.4 and 3.5, two special cases for Amplitude to Amplitude (AM/AM) and Amplitude to Phase (AM/PM) estimation that are designed to circumvent these problems are discussed in depth. The chapter is concluded in section 3.6 with a summary of the most important results.

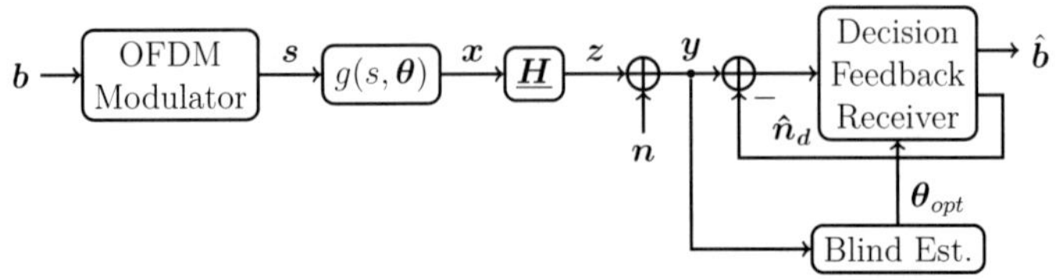

Figure 3.1.: System model for blind feed forward estimation

3.2. System Model

The system model is shown in Figure 3.1. The system is modeled entirely in the complex baseband, hence, all time domain samples are complex valued. The transmitter part is an OFDM modulator that transforms a vector $\boldsymbol{b}$ of bits into a zero-mean complex normally distributed time domain signal $\boldsymbol{s}$ so that

$$\boldsymbol{s} \sim CN(\mathbf{0}, \underline{\boldsymbol{C}}_s), \tag{3.1}$$

where $\underline{\boldsymbol{C}}_s$ is the covariance matrix of $\boldsymbol{s}$. This signal is then subject to a memoryless nonlinear distortion $g(s, \boldsymbol{\theta})$ which represents the nonlinear effects of the transmit amplifier. For this nonlinearity it is assumed that the characteristic is known at the receiver but it depends on a vector of parameters $\boldsymbol{\theta}$ that is unknown. An example would be to assume an Solid State Power Amplifier (SSPA) nonlinearity for which $\boldsymbol{\theta} = [A_{max}, p]$. Afterwards, the signal is subject to frequency selective fading represented by the circular convolution matrix $\underline{\boldsymbol{H}}$ which is assumed as perfectly known in the receiver (some examples where this is not the case are shown in chapter 5). This kind of fading could for example stem from multipath propagation or single frequency networks in broadcasting. By setting $\underline{\boldsymbol{H}} = h_0 \cdot \underline{\boldsymbol{I}}$, frequency flat fading can be represented where h_0 is the instantaenous complex fading coefficient for this OFDM symbol. Finally, for $h_0 = 1$, there is no fading and the model represents an Additive White Gaussian Noise (AWGN) channel. Afterwards, the signal is superimposed with AWGN represented by the vector $\boldsymbol{n}$ for which

$$\boldsymbol{n} \sim CN(\mathbf{0}, \sigma_n^2 \underline{\boldsymbol{I}}) \tag{3.2}$$

holds. This models background noise effects which for example stem from thermal noise in the transmitter and receiver circuitry. The resulting signal $\boldsymbol{y}$ is now received by the decision feedback receiver as described in section 2.3.3. It is also fed into a blind estimator which outputs an estimate on the parameter vector $\boldsymbol{\theta}_{opt}$ which is used for nonlinearity mitigation in the receiver.

In this scenario, the blind estimator only has the received signal to work with. This makes the estimator mostly independent of the remaining receiver structure so that other architectures could be used as well. Furthermore, for estimation it is not even required that an OFDM signal is used as long as (3.1) holds for the transmitted signal. In the following, the novel blind estimation procedure is described.

3.3. Blind Maximum Likelihood Estimator

The Maximum Likelihood Estimator is often used to obtain practical estimators. While other estimators might exhibit certain desirable properties, they are often not existing or difficult to derive for complicated practical problems. The Maximum Likelihood Estimator (MLE) has the advantage of being a "turn-the-crank" procedure that can be implemented for many complicated problems either analytically or numerically. Additionally, important properties are proven to hold for large observation sizes. For those reasons, the MLE forms the basis for solving many practical estimation problems.

This section gives a short overview of the MLE. A good introduction can be found in [Kay93a].

3.3.1. Definition

Let $\boldsymbol{\zeta}$ be an observation of some random variable $\boldsymbol{Z}$, for example a measurement of some nondeterministic system output. The random variable depends on a parameter vector $\boldsymbol{\theta}$ so that its Probability Density Function (PDF) can be written as $p_\zeta(\boldsymbol{\zeta} \mid \boldsymbol{\theta})$ where "|" denotes that this is a conditional PDF of $\boldsymbol{\zeta}$ given $\boldsymbol{\theta}$. The parameters are deterministic and time-invariant but unknown to the observer and to be estimated only from the observations. The maximum likelihood estimate $\boldsymbol{\theta}_{opt}$ of the parameters is given as:

$$\boldsymbol{\theta}_{opt} = \arg\max_{\boldsymbol{\theta}} \left[p_\zeta(\boldsymbol{\zeta} \mid \boldsymbol{\theta}) \right] . \tag{3.3}$$

$p_\zeta(\boldsymbol{\zeta} \mid \boldsymbol{\theta})$ can be interpreted as proportional to the probability that an observation of $\boldsymbol{Z}$ is in a very small volume around $\boldsymbol{\zeta}$ and hence, how likely it is to observe $\boldsymbol{\zeta}$ itself under the parameters $\boldsymbol{\theta}$. If its value is small, observing $\boldsymbol{\zeta}$ under the parameters $\boldsymbol{\theta}$ is not very likely. On the other hand, for large values of $p_\zeta(\boldsymbol{\zeta} \mid \boldsymbol{\theta})$ it is more likely to observe $\boldsymbol{\zeta}$ under the parameters $\boldsymbol{\theta}$. For a given observation $\boldsymbol{\zeta}$, the Maximum Likelihood (ML)-estimate $\boldsymbol{\theta}_{opt}$ is the one achieving the largest value of the PDF and hence having the highest likelihood of causing the observation $\boldsymbol{\zeta}$. Summarized in a single sentence:

> *The Maximum Likelihood Estimate consists of those parameters $\boldsymbol{\theta}_{opt}$ that make the observations $\boldsymbol{\zeta}$ most plausible.*

Care must be taken not to misinterpret this statement. The estimate $\boldsymbol{\theta}_{opt}$ is not the parameter vector that has the highest probability and $p_\zeta(\boldsymbol{\zeta} \mid \boldsymbol{\theta})$ is not some form of probability density function for $\boldsymbol{\theta}$. In fact, since $\boldsymbol{\theta}$ is deterministic, it does not have a density function. To distinguish this, $p_\zeta(\boldsymbol{\zeta} \mid \boldsymbol{\theta})$ is often called the likelihood function of $\boldsymbol{\theta}$ when $\boldsymbol{\zeta}$ is fixed and $\boldsymbol{\theta}$ is variable.

In many applications, there exist multiple mutually independent observations $\boldsymbol{\zeta}_k$. The MLE in this case is given as:

$$\boldsymbol{\theta}_{opt} = \arg\max_{\boldsymbol{\theta}} \left[\prod_k p_{\zeta_k}(\boldsymbol{\zeta}_k \mid \boldsymbol{\theta}) \right] . \tag{3.4}$$

For reasons of numerical stability, the logarithm of the likelihood function is often used:

$$\boldsymbol{\theta}_{opt} = \arg\max_{\boldsymbol{\theta}} \left[\sum_k \log\, p_{\zeta_k}(\boldsymbol{\zeta}_k \mid \boldsymbol{\theta}) \right] = \arg\max_{\boldsymbol{\theta}} \left[\sum_k l_{\zeta_k}(\boldsymbol{\zeta}_k \mid \boldsymbol{\theta}) \right] \tag{3.5}$$

and $l_{\zeta_k}(\boldsymbol{\zeta}_k \mid \boldsymbol{\theta})$ is called the log-likelihood function. This is possible because the logarithm is strictly monotonic and hence, the logarithm of a function will have the same extrema as the function itself. In the following, only the MLE for one observation is derived. Multiple observations can be combined as shown in (3.4).

Properties A very important property of the MLE is *asymptotic efficiency*. For a large amount of observations $N_o \to \infty$, the estimator attains the Cramér-Rao Lower Bound (CRLB) which is a bound on the minimum achievable error variance of any unbiased estimator. This topic is covered in depth later. In principle this implies that the estimation of $\boldsymbol{\theta}$ can be done with arbitrarily high precision by increasing the amount of independent observations.

The estimator is also *asymptotically unbiased*. This means that if the amount of observations $N_o \to \infty$, the expectation of the estimate $\mathrm{E}\{\boldsymbol{\theta}_{opt}\} \to \boldsymbol{\theta}$. For an unbiased estimator, subsequent estimates can be averaged to obtain a more exact estimate.

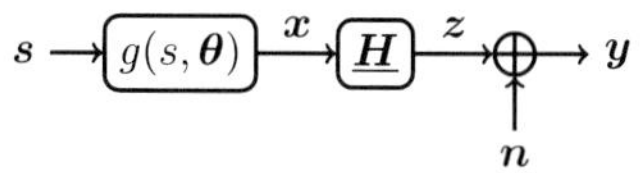

Figure 3.2.: Estimation Model

The MLE has been studied and described extensively in the literature. For example, a more elaborated introduction is given in [Kay93b].

3.3.2. Application to nonlinearity estimation

From the definition of the MLE follow three main requirements that need to be fulfilled so that it can be used:

1. The parameters $\boldsymbol{\theta}$ are deterministic and time-invariant
2. Observations can be made from the system
3. The observations $\boldsymbol{\zeta}$ depend on the parameters and $p_{\zeta}(\boldsymbol{\zeta} \mid \boldsymbol{\theta})$ can be evaluated

The first point is not true for the system under consideration since the nonlinearity parameters can change for a number of reasons such as component aging or a change in device temperature. However, all of these changes usually happen very slowly. It will be shown later that the required amount of observations for an accurate estimate is rather small. Therefore the parameters can be considered as constant for one set of observations. It might be required to repeat the estimation periodically, though.

The second point is obvious. The observations are the received time-samples $\boldsymbol{y}$ belonging to the OFDM symbols. Given frequent transmissions, the amount of observations available for estimation can also be controlled almost arbitrarily given enough memory and computational power.

The third point is problematic. The received samples are influenced by the parameters. However, the conditional distribution $p_{\zeta}(\boldsymbol{\zeta} \mid \boldsymbol{\theta})$ is not known yet. Since its derivation is not trivial, it is handled in detail in the next section.

3.3.3. Distribution of received samples

Figure 3.2 shows the part of the system model that is relevant for the intended parameter estimation. Starting from a known distribution $p_s(\boldsymbol{s})$, the distribution $p_y(\boldsymbol{y} \mid \boldsymbol{\theta})$ shall now be derived. The random variable transformation technique is extensively used in the following. An introduction is given in appendix A.

Distribution before nonlinearity The $[N_c \times 1]$-vector $\boldsymbol{s}$ contains the OFDM time domain signal before nonlinear distortion. It has been discussed already that it exhibits a zero-mean complex normal distribution with covariance matrix $\underline{\boldsymbol{C}}_s$, i.e.,:

$$\boldsymbol{s} \sim CN(\boldsymbol{0}, \underline{\boldsymbol{C}}_s) \tag{3.6}$$

$$p_{\boldsymbol{s}}(\boldsymbol{s}) = \frac{1}{\pi^{N_c} \left|\underline{\boldsymbol{C}}_s\right|} \exp\left(-\boldsymbol{s}^H \underline{\boldsymbol{C}}_s^{-1} \boldsymbol{s}\right) . \tag{3.7}$$

When the average power σ_s^2 of all subcarriers is equal, this is simplified to:

$$\boldsymbol{s} \sim CN(\boldsymbol{0}, \sigma_s^2 \underline{\boldsymbol{I}}_{N_c}) \tag{3.8}$$

$$p_{\boldsymbol{s}}(\boldsymbol{s}) = \prod_{k=0}^{N_c-1} \frac{1}{\pi\sigma_s^2} \exp\left(-\frac{|s_k|^2}{\sigma_s^2}\right) , \tag{3.9}$$

where $\underline{\boldsymbol{I}}_{N_c}$ denotes the $[N_c \times N_c]$-identity matrix. However, the carriers at the edges of the frequency band as well as around the DC component are often unused. Those carriers are called guardband carriers. Furthermore, in many common technologies, e.g. water-filling [JK03, RC00], each subcarrier is assigned a different power level. If there are guardband carriers or per-subcarrier power allocation is used this can be modeled using a $[N_c \times 1]$-vector $\boldsymbol{H_p}$. The l-th element of this vector contains a power scaling factor for the l-th subcarrier. This is similar to frequency selective filtering of white Gaussian noise and the circular convolution matrix $\underline{\boldsymbol{H}}_p$ is given as

$$\underline{\boldsymbol{H}}_p = \frac{\underline{\boldsymbol{W}}\, \mathrm{diag}(\boldsymbol{H_p}) \underline{\boldsymbol{W}}^H}{\sqrt{N_c}} , \tag{3.10}$$

where $\underline{\boldsymbol{W}}$ is the $[N_c \times N_c]$ Discrete time Fourier Transformation (DFT)-matrix defined as follows:

$$\underline{\boldsymbol{W}} = \left(\frac{\omega^{lm}}{\sqrt{N_c}}\right)_{l,m=[0\ldots N_c-1]} \quad \text{with } \omega = \exp\left(\frac{j2\pi}{N_c}\right) . \tag{3.11}$$

The covariance matrix of the resulting colored noise is then given as $\underline{\boldsymbol{C}}_s = \underline{\boldsymbol{H}}_p \underline{\boldsymbol{H}}_p^H$.

Hence, the following estimator can be applied to any signal exhibiting a complex normal distribution in time-domain as long as it can be characterized as in (3.6). This covers OFDM signals with per-subcarrier power allocation and guardband carriers. Some practical problems are covered in the next section.

Transformation to polar coordinates Since the memoryless nonlinearity is usually defined in terms of amplitude and phase distortion it is useful to transform the random vector $\boldsymbol{s}$ into polar coordinates. Splitting $\boldsymbol{s}$ up into the complex components $\boldsymbol{s}_r = \Re\{\boldsymbol{s}\}$ and $\boldsymbol{s}_i = \Im\{\boldsymbol{s}\}$, the coordinate transformation is given by:

$$\begin{aligned} \boldsymbol{s}_A &= \sqrt{\boldsymbol{s}_r^2 + \boldsymbol{s}_i^2} \\ \boldsymbol{s}_\phi &= \arg(\boldsymbol{s}) \end{aligned} \tag{3.12}$$

with all operations being element-wise. To calculate the random variable transformation, the inverse of (3.12) is required:

$$\begin{aligned} \boldsymbol{s}_r &= \boldsymbol{s}_A \cdot \cos \boldsymbol{s}_\phi \\ \boldsymbol{s}_i &= \boldsymbol{s}_A \cdot \sin \boldsymbol{s}_\phi \,. \end{aligned} \tag{3.13}$$

J is the determinant of the Jacobian matrix which consists of all first partial derivatives of the transformation. In our case it can be written as follows:

$$J(\boldsymbol{s}_r, \boldsymbol{s}_i) = \begin{vmatrix} \underline{\boldsymbol{A}} & \underline{\boldsymbol{B}} \\ \underline{\boldsymbol{C}} & \underline{\boldsymbol{D}} \end{vmatrix}^{-1} \tag{3.14}$$

with

$$\begin{aligned} \underline{\boldsymbol{A}} &= \operatorname{diag}(\frac{\partial \boldsymbol{s}_r}{\partial \boldsymbol{s}_A}) = \operatorname{diag}(\cos \boldsymbol{s}_\phi) \\ \underline{\boldsymbol{B}} &= \operatorname{diag}(\frac{\partial \boldsymbol{s}_r}{\partial \boldsymbol{s}_\phi}) = \operatorname{diag}(-\boldsymbol{s}_A \cdot \sin \boldsymbol{s}_\phi) \\ \underline{\boldsymbol{C}} &= \operatorname{diag}(\frac{\partial \boldsymbol{s}_i}{\partial \boldsymbol{s}_A}) = \operatorname{diag}(\sin \boldsymbol{s}_\phi) \\ \underline{\boldsymbol{D}} &= \operatorname{diag}(\frac{\partial \boldsymbol{s}_i}{\partial \boldsymbol{s}_\phi}) = \operatorname{diag}(\boldsymbol{s}_A \cdot \cos \boldsymbol{s}_\phi)\,. \end{aligned} \tag{3.15}$$

The transformation is memoryless so all partial derivatives with unequal time-indices are zero yielding the four diagonal matrices as shown above. Then,

$$J(\boldsymbol{s}_r, \boldsymbol{s}_i) = \left|\underline{\boldsymbol{AD}} - \underline{\boldsymbol{BC}}\right|^{-1} = \left|\operatorname{diag}(\boldsymbol{s}_A)\right|^{-1} = \frac{1}{\prod_{k=0}^{N_c-1} [\boldsymbol{s}_A]_k}\,. \tag{3.16}$$

Finally, using the rules for random variable transformation [Pap91], the PDF $p_{\boldsymbol{s}_{A/\phi}}(\boldsymbol{s}_A, \boldsymbol{s}_\phi)$ of the vectors in polar coordinates is given in terms of the original PDF

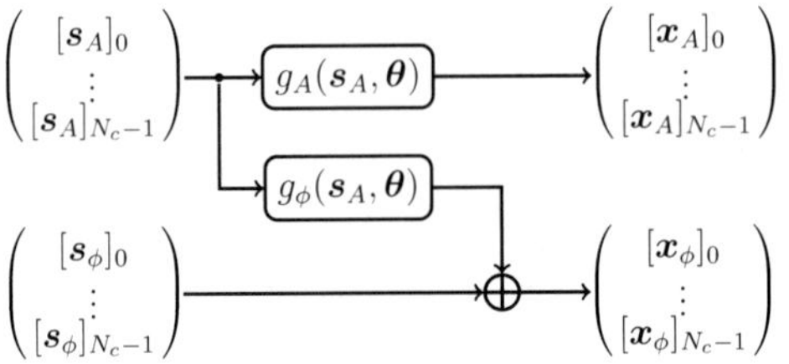

Figure 3.3.: Random variable transformation of the memoryless nonlinearity

$p_s(\boldsymbol{s})$ as:

$$p_{\boldsymbol{s}_{A/\phi}}(\boldsymbol{s}_A, \boldsymbol{s}_\phi) = \frac{p_s\left(\boldsymbol{s}_r(\boldsymbol{s}_A, \boldsymbol{s}_\phi) + j\boldsymbol{s}_i(\boldsymbol{s}_A, \boldsymbol{s}_\phi)\right)}{J(\boldsymbol{s}_r, \boldsymbol{s}_i)} = p_s\left(\boldsymbol{s}_A \cdot \cos \boldsymbol{s}_\phi + j\boldsymbol{s}_A \cdot \sin \boldsymbol{s}_\phi\right) \cdot \prod_{k=0}^{N_c-1} [\boldsymbol{s}_A]_k \, . \tag{3.17}$$

Distribution after Memoryless Nonlinearity Again, using the random variable transformation technique, the PDF of $\boldsymbol{x}$ after the memoryless nonlinearity is calculated. Since this transformation is done separately for each time sample, all following vector operations are element-wise. Figure 3.3 shows how the inputs and outputs of the transformation are related. The output $\boldsymbol{x}$ in polar coordinates is given as:

$$\begin{aligned} \boldsymbol{x}_A &= g_A(\boldsymbol{s}_A, \boldsymbol{\theta}) \\ \boldsymbol{x}_\phi &= \boldsymbol{s}_\phi + g_\phi(\boldsymbol{s}_A, \boldsymbol{\theta}) \, . \end{aligned} \tag{3.18}$$

Again, for the random variable transformation, the inverse of (3.18) is required. It is given as:

$$\begin{aligned} \boldsymbol{s}_A &= g_A^{-1}(\boldsymbol{x}_A, \boldsymbol{\theta}) \\ \boldsymbol{s}_\phi &= \boldsymbol{x}_\phi - g_\phi(\boldsymbol{s}_A, \boldsymbol{\theta}) \, . \end{aligned} \tag{3.19}$$

Only the inverse of the AM/AM distortion is required. The Jacobian of this transformation is constructed in the same way as for the rectangular-to-polar transformation and consists of four diagonal matrices as follows:

$$\underline{\boldsymbol{A}} = \operatorname{diag}\left(\frac{\partial \boldsymbol{x}_A}{\partial \boldsymbol{s}_A}\right) = \operatorname{diag}\left(\frac{\partial g_A(\boldsymbol{s}_A, \boldsymbol{\theta})}{\partial \boldsymbol{s}_A}\right) \tag{3.20}$$

$$\underline{\boldsymbol{B}} = \operatorname{diag}\left(\frac{\partial \boldsymbol{x}_A}{\partial \boldsymbol{s}_\phi}\right) = \underline{\mathbf{0}} \tag{3.21}$$

$$\underline{\boldsymbol{C}} = \operatorname{diag}\left(\frac{\partial \boldsymbol{x}_\phi}{\partial \boldsymbol{s}_A}\right) = \operatorname{diag}\left(\frac{\partial g_\phi(\boldsymbol{s}_A, \boldsymbol{\theta})}{\partial \boldsymbol{s}_A}\right) \tag{3.22}$$

$$\underline{\boldsymbol{D}} = \operatorname{diag}\left(\frac{\partial \boldsymbol{x}_\phi}{\partial \boldsymbol{s}_\phi}\right) = \underline{\boldsymbol{I}}_{N_c} . \tag{3.23}$$

The Jacobian determinant is then given as:

$$J = \begin{vmatrix} \underline{\boldsymbol{A}} & \underline{\boldsymbol{B}} \\ \underline{\boldsymbol{C}} & \underline{\boldsymbol{D}} \end{vmatrix} = |\underline{\boldsymbol{A}}\underline{\boldsymbol{D}} - \underline{\boldsymbol{B}}\underline{\boldsymbol{C}}| = |\underline{\boldsymbol{A}}| = \prod_{k=0}^{N_c-1} g'_A(g_A^{-1}(\boldsymbol{x}_A, \boldsymbol{\theta}), \boldsymbol{\theta}) \tag{3.24}$$

with $g'_A(x, \boldsymbol{\theta}) = \frac{\partial g_A(x,\boldsymbol{\theta})}{\partial x}$. Putting everything together, the joint amplitude and phase distribution after the nonlinearity is given as:

$$p_{\boldsymbol{x}_{A/\phi}}(\boldsymbol{x}_A, \boldsymbol{x}_\phi \mid \boldsymbol{\theta}) = \frac{p_{\boldsymbol{s}_{A/\phi}}\left(g_A^{-1}(\boldsymbol{x}_A, \boldsymbol{\theta}), \boldsymbol{x}_\phi - g_\phi(g_A^{-1}(\boldsymbol{x}_A, \boldsymbol{\theta}), \boldsymbol{\theta})\right)}{\prod_{k=0}^{N_c-1} g'_A(g_A^{-1}([\boldsymbol{x}_A]_k, \boldsymbol{\theta}), \boldsymbol{\theta})} . \tag{3.25}$$

To calculate this distribution, the first derivative of the AM/AM characteristic must exist and g_A has to be invertible. The first derivative is available for many commonly used nonlinearity models. However, many AM/AM models are not invertible. For example, the AM/AM characteristic of the Saleh model is not bijective and has two possible input amplitudes for every output amplitude.

Non-bijective AM/AM characteristic The solution to this issue lies in the interpretation of the variable transformation. Loosely spoken, the probability of a certain system output is equal to the probability of the system input that produces this response. If the system characteristic is not bijective, multiple inputs can cause a single output. Therefore, the probability for such an output is the sum of all input probabilities that result in this output.

If the amount of solutions is finite (i.e., $g_A(\boldsymbol{s}, \boldsymbol{\theta})$ is piecewise monotonic), the resulting PDF can be determined. Let $\boldsymbol{x}_A = g_A(\boldsymbol{s}_A, \boldsymbol{\theta})$ have V solutions and the v-th solution is given by $\boldsymbol{s}_A^{(v)} = g_A^{(v)}(\boldsymbol{x}_A, \boldsymbol{\theta})$ so that $\boldsymbol{x}_A = g_A(\boldsymbol{s}_A^{(v)}, \boldsymbol{\theta})$ for all v. For a non-bijective AM/AM characteristic, (3.25) is given as:

$$p_{\boldsymbol{x}_{A/\phi}}(\boldsymbol{x}_A, \boldsymbol{x}_\phi \mid \boldsymbol{\theta}) = \sum_{v=0}^{V-1} \frac{p_{\boldsymbol{s}_{A/\phi}}\left(g_A^{(v)}(\boldsymbol{x}_A, \boldsymbol{\theta}), \boldsymbol{x}_\phi - g_\phi(g_A^{(v)}(\boldsymbol{x}_A, \boldsymbol{\theta}), \boldsymbol{\theta})\right)}{\prod_{k=0}^{N_c-1} g'_A(g_A^{(v)}([\boldsymbol{x}_A]_k, \boldsymbol{\theta}), \boldsymbol{\theta})} . \tag{3.26}$$

This is equal to (3.25) for $v = 1$ and $g_A^{(0)} = g_A^{-1}$. Using this, most models for nonlinear amplifiers and other nonlinearities can be handled by this framework. Some caveats of this approach are described in the next section. Still not covered are nonlinearities such as the soft limiter or quantizers where an infinite amount of inputs leads to a single output. This results in a mixed or purely discrete distribution and describing this in a general way is very involved. However, for a concrete characteristic the post-nonlinearity PDF can usually still be calculated and the rest of the framework can handle it.

Transformation to carthesian coordinates For the remaining derivation it is favorable to work with carthesian coordinates. The inverse is exactly the opposite of the earlier transformation to polar coordinates. For brevity, the final result is:

$$p_{\boldsymbol{x}}(\boldsymbol{x} \mid \boldsymbol{\theta}) = \frac{p_{\boldsymbol{x}_{A/\phi}}(|\boldsymbol{x}|, \arg \boldsymbol{x} \mid \boldsymbol{\theta})}{\prod_{k=0}^{N_c-1} |x_k|} . \tag{3.27}$$

Distribution after frequency selective channel The frequency selective channel is represented by the matrix $\underline{\boldsymbol{H}}$. The signal $\boldsymbol{z}$ at the output of the channel is given by:

$$\boldsymbol{z} = \underline{\boldsymbol{H}} \boldsymbol{x} \tag{3.28}$$

and the inverse by:

$$\boldsymbol{x} = \underline{\boldsymbol{H}}^{-1} \boldsymbol{z} . \tag{3.29}$$

Since (3.28) is a linear transformation, it can be shown that the Jacobian matrix is $\underline{\boldsymbol{H}}$ and therefore the Jacobian determinant is given by:

$$J = |\underline{\boldsymbol{H}}| . \tag{3.30}$$

Using this, the PDF of $\boldsymbol{z}$ is given by:

$$p_{\boldsymbol{z}}(\boldsymbol{z} \mid \boldsymbol{\theta}) = \frac{p_{\boldsymbol{x}}(\underline{\boldsymbol{H}}^{-1} \boldsymbol{z} \mid \boldsymbol{\theta})}{|\underline{\boldsymbol{H}}|} . \tag{3.31}$$

Some issues concerning deep fades, i.e., strongly attenuated subcarriers due to deep notches in the channel transfer function, are adressed in the next section.

Distribution after AWGN The received signal $\boldsymbol{y}$ that forms the observations of the ML-estimator consists of the sum of the distorted source signal $\boldsymbol{z}$ and the AWGN vector $\boldsymbol{n}$:

$$\boldsymbol{y} = \boldsymbol{z} + \boldsymbol{n}\,. \tag{3.32}$$

The noise is usually induced by the receiver circuitry and commonly assumed to be independent of the transmitted signal. The PDF of a sum of independent random variables is given by the convolution of the individual PDFs [Pap91]:

$$p_{\boldsymbol{y}}(\boldsymbol{y} \mid \boldsymbol{\theta}) = p_{\boldsymbol{z}}(\boldsymbol{z} \mid \boldsymbol{\theta}) * p_{\boldsymbol{n}}(\boldsymbol{n}) = \int_{\mathbb{C}^{N_c}} \cdots \int p_{\boldsymbol{z}}(\boldsymbol{y}' \mid \boldsymbol{\theta}) \cdot p_{\boldsymbol{n}}(\boldsymbol{y} - \boldsymbol{y}')\mathrm{d}\boldsymbol{y}'\,. \tag{3.33}$$

Conclusions In order to use the maximum likelihood technique to estimate the nonlinearity parameters $\boldsymbol{\theta}$ the density function $p_{\boldsymbol{y}}(\boldsymbol{y} \mid \boldsymbol{\theta})$ of the observations conditioned on the parameters needs to be known. This density has been derived in this section and is now summarized. Given an observation vector $\boldsymbol{y}$ and parameters $\boldsymbol{\theta}$, the density function can be computed as follows:

$$p_{\boldsymbol{y}}(\boldsymbol{y} \mid \boldsymbol{\theta}) = \int_{\mathbb{C}^{N_c}} \cdots \int p_{\boldsymbol{z}}(\boldsymbol{y}' \mid \boldsymbol{\theta}) \cdot p_{\boldsymbol{n}}(\boldsymbol{y} - \boldsymbol{y}')\mathrm{d}\boldsymbol{y}' \tag{3.34}$$

$$p_{\boldsymbol{z}}(\boldsymbol{z} \mid \boldsymbol{\theta}) = \frac{p_{\boldsymbol{x}}(\underline{\boldsymbol{H}}^{-1}\boldsymbol{z} \mid \boldsymbol{\theta})}{|\underline{\boldsymbol{H}}|} \tag{3.35}$$

$$p_{\boldsymbol{x}}(\boldsymbol{x} \mid \boldsymbol{\theta}) = \frac{p_{\boldsymbol{x}_{A/\phi}}(|\boldsymbol{x}|, \arg \boldsymbol{x} \mid \boldsymbol{\theta})}{\prod_{k=0}^{N_c-1} |x_k|} \tag{3.36}$$

$$p_{\boldsymbol{x}_{A/\phi}}(\boldsymbol{x}_A, \boldsymbol{x}_\phi \mid \boldsymbol{\theta}) = \frac{p_{\boldsymbol{s}_{A/\phi}}\left(g_A^{-1}(\boldsymbol{x}_A, \boldsymbol{\theta}), \boldsymbol{x}_\phi - g_\phi(g_A^{-1}(\boldsymbol{x}_A, \boldsymbol{\theta}), \boldsymbol{\theta})\right)}{\prod_{k=0}^{N_c-1} g_A'(g_A^{-1}([\boldsymbol{x}_A]_k, \boldsymbol{\theta}), \boldsymbol{\theta})} \tag{3.37}$$

$$p_{\boldsymbol{s}_{A/\phi}}(\boldsymbol{s}_A, \boldsymbol{s}_\phi) = p_{\boldsymbol{s}}\left(\boldsymbol{s}_A \cdot \cos \boldsymbol{s}_\phi + j\boldsymbol{s}_A \cdot \sin \boldsymbol{s}_\phi\right) \cdot \prod_{k=0}^{N_c-1} [\boldsymbol{s}_A]_k \tag{3.38}$$

$$p_{\boldsymbol{s}}(\boldsymbol{s}) = \frac{1}{\pi^{N_c}|\underline{\boldsymbol{C}}_s|} \exp\left(\boldsymbol{s}^H \underline{\boldsymbol{C}}_s^{-1} \boldsymbol{s}\right) \tag{3.39}$$

Despite the addition of the noise, all steps are deterministic random variable transformations. As such, the density function of the output is represented in terms of the density function of the respective input.

In this section, the PDF for a nonlinearly distorted, possibly oversampled OFDM time signal in a frequency selective environment has been derived. For practical applications, there remain some issues which are inspected now.

3.3.4. Practical Issues

While mathematically correct, there are three practical issues when evaluating the density function of the observed samples: High dimensional convolution, ill-conditioned matrices and non-invertible AM/AM characteristics.

High dimensional convolution From the experience of the author, it is usually possible to derive the density function before AWGN analytically, given the frequency selective channel in (3.31) is known. However, finding the analytical result of the following convolution has shown to be prohibitively difficult for the common nonlinearity models presented in section 2.2.2 so that the calculation can only be done numerically.

The dimension of the convolution is twice the amount of subcarriers[1]. It was mentioned in 2.1.4 that $N_c \geq 128$ in order for the Gaussian assumption to hold and modern OFDM systems can easily have thousands of subcarriers. A highly-dimensional numerical convolution suffers from both high computational complexity and large memory requirements. When sampling the PDFs with L samples per dimension, memory to store L^{2N_c} numbers is required[2]. Even for $N_c = 128$ and $L = 2$ this is completely out of reach for any computer system today.

To calculate the density function for one observation $\boldsymbol{y}$, only one point of the whole convolution is required. The amount of multiplications required for this is L^{4N_c}.[3] While there are efficient algorithms that use redundancies when calculating the full convolution and therefore reduce the amount of multiplications per output sample, the minimum number of multiplications to calculate at least one output sample cannot be lower. It is again easy to see that the amount of multiplications required is too high to be feasible.

This problem is addressed by assuming a lower dimensionality at the cost of estimation performance as shown in sections 3.4.5 and 3.5.4.

[1] Factor two due to complex convolution
[2] Factor two due to complex convolution
[3] Factor 4 due to complex numbers

Non-invertible AM/AM characteristic It has been shown in (3.26) how non-bijective AM/AM characteristics can be handled. While this is a mathematically sound solution it bears a subtle problem when practically implemented. Consider the following toy example:

Let $g_A([\boldsymbol{s}_A]_k)$ be a non-bijective memoryless nonlinearity and $g_A(1) = g_A(2) = 4$. The vector $\boldsymbol{x} = [4, 4]$ is observed at the output of the nonlinearity. There are $V = 4$ solutions to the problem $\boldsymbol{x} = g_A(\boldsymbol{s})$, namely $\boldsymbol{s}^{(0)} = [1, 1]$, $\boldsymbol{s}^{(1)} = [1, 2]$, $\boldsymbol{s}^{(2)} = [2, 1]$ and $\boldsymbol{s}^{(3)} = [2, 2]$.

It can be shown that the amount of solutions is $V = (V_g)^{N_c}$ if the nonlinearity characteristic is not bijective and every output sample maps to V_g input samples. For example, for the AM/AM characteristic of the Saleh model, each output sample has two possible input samples. This results in 2^{N_c} solutions and the same amount of summation terms in (3.26). Again, calculating this for concrete observations in a practical scenario is hardly feasible so that this general solution is effectively limited to bijective AM/AM characteristics in practice.

This problem is alleviated as well by the dimensionality reduction mentioned earlier.

Ill-conditioned matrices (3.10) describes that the circular convolution matrix $\underline{\boldsymbol{H}}_p = N_c^{-1/2}\underline{\boldsymbol{W}}\,\mathrm{diag}(\boldsymbol{H_p})\underline{\boldsymbol{W}}^H$ with the frequency response vector $\boldsymbol{H_p}$ and DFT-matrix $\underline{\boldsymbol{W}}$. By definition, the DFT-matrix is unitary and normal, i.e., $\underline{\boldsymbol{W}}^{-1} = \underline{\boldsymbol{W}}^H$ and $\underline{\boldsymbol{W}}^H\underline{\boldsymbol{W}} = \underline{\boldsymbol{W}\boldsymbol{W}}^H$. It can be shown that the Eigenvalues of $\underline{\boldsymbol{H}}_p$ are given by $N_c^{-1/2}\boldsymbol{H_p}$. In other words, the scaling factor applied to each subcarrier is also an Eigenvalue of the convolution matrix. As a result, as soon as one subcarrier is scaled to zero, the circular convolution matrix becomes singular.

This is problematic especially when estimating nonlinearity parameters in an OFDM system with guardband carriers. Those carriers, per definition, do not have any power allocated. When modelling the system as in section 3.3.3, this will lead to a singular circular convolution matrix. Consequently, the covariance matrix of the distribution of the modulated OFDM time signal also becomes singular. The result is a degenerated multivariate normal distribution that does not have a density function and hence cannot be estimated using a maximum likelihood estimator. This can also happen when the frequency selective channel exhibits deep fades. Due to limited numerical accuracy of the hardware, these can come out as zero in the channel estimation resulting in a singular convolution matrix. This renders (3.31) unsolvable and therefore the density function of the received samples does not exist.

The problem can be alleviated by assuming the carriers to have a very small amount of power. Hence, instead of setting the elements of $\underline{\boldsymbol{H}}_p$ that represent the guardband carriers to zero, they are being set to a small nonzero value. Especially in the case of guardband carriers, this introduces a systematic error into the estimation problem. Furthermore, while no longer singular, the matrices become near-singular or ill-conditioned. The result is that the estimator becomes very sensitive to small numerical errors in the inputs that arise due to the limited accuracy of the underlying computing platform. For a circular convolution matrix $\underline{\boldsymbol{H}}_p$, the condition number $\kappa(\underline{\boldsymbol{H}}_p)$ is the ratio of the largest and smallest absolute Eigenvalue, i.e.,:

$$\kappa(\underline{\boldsymbol{H}}_p) = \frac{\max|\boldsymbol{H}_p|}{\min|\boldsymbol{H}_p|} . \tag{3.40}$$

As a general rule of thumb, for $\kappa(\underline{\boldsymbol{H}}_p) = 10^q$, q digits of precision are lost every time a vector is multiplied with the inverse of the ill-conditioned matrix [CK07]. Under this conditions, setting zeroed carriers to a small value becomes a tradeoff between error induced by falsely assuming the carriers to be nonzero and numerical stability.

3.3.5. Conclusions

The maximum likelihood estimator for the parameter vector $\boldsymbol{\theta}$ of a memoryless nonlinearity exhibiting AM/AM and AM/PM distortions has been derived. It can be applied to white and colored noise input processes subject to nonlinear distortion followed by frequency selective fading and superimposed by additive white Gaussian noise. In practical implementations, a highly-dimensional convolution or non-bijective amplitude distortion can lead to an unreasonably high computational effort. Furthermore, the method is prone to numerical instability in certain circumstances. In the following, two special cases are presented where these issues are alleviated and that can still be applied to a wide range of problems.

3.4. Blind ML-Estimator for AM/AM nonlinearities in flat fading

The first special case is an OFDM system without guardband and a memoryless nonlinearity that exhibits no phase distortion in a frequency flat fading scenario

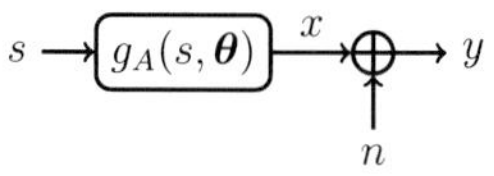

Figure 3.4.: Simplified signal model for estimation

[DF11].

3.4.1. Motivation

A major hindrance when using the general estimator is correlated samples. This leads to highly-dimensional convolutions as well as numerical instability when carriers are close to zero power. When assuming an OFDM system without guardband carriers and with equal power per subcarrier and flat fading, both problems disappear. Several techniques are discussed later that allow the application to OFDM systems with guardband and in frequency selective conditions.

Limiting the estimation to amplitude distortions is not a design decision but a consequence of independent time domain samples. It will be shown that under this condition, blindly estimating the phase characteristic is not possible. However, many amplifiers are designed to exhibit very low AM/PM distortion. For example, the SSPA-model was specifically designed to resemble modern solid state amplifiers and does not exhibit phase distortions. Therefore, assuming only an AM/AM distortion is valid. It can also be seen as the first of a two-step estimation with the phase distortion characteristic estimated in a subsequent step.

3.4.2. Derivation of the conditional density

Figure 3.4 shows the simplified signal model used for parameter estimation. Again, the conditional distribution of the observations $p_y(y \mid \boldsymbol{\theta})$ is derived in order to formulate the maximum likelihood estimator. If not mentioned otherwise, all of the following applies independently for each time sample. A symbol without bold vector notation refers to a sample at a certain arbitrary time.

This chapter uses the theory of random variable transformation extensively. A short introduction is given in Appendix A and a good reference is [Pap91].

Distribution before nonlinearity Assume the OFDM signal is transmitted without any guardband carriers and with equal power per subcarrier. As mentioned in

(3.9), the resulting time domain samples have an i.i.d. zero-mean complex normal distribution, i.e.,

$$s \sim CN(0, \sigma_s^2)\,. \tag{3.41}$$

It is well known that the density function is given by

$$p_s(s) = \frac{1}{\pi\sigma_s^2} \exp\left(-\frac{ss^*}{\sigma_s^2}\right)\,. \tag{3.42}$$

It is notable that the density function exhibits circular symmetry, i.e., the value of the PDF only depends on the amplitude $|s|^2 = ss^*$ and not on the phase of the time sample.

Transformation to polar coordinates Again it is useful to transform the random variable to polar coordinates before considering the nonlinear distortion. Applying (3.17) to the scalar case leads to

$$p_{s_{A/\phi}}(s_A, s_\phi) = s_A \cdot p_s\left(s_A e^{js_\phi}\right)\,. \tag{3.43}$$

Distribution after nonlinearity Similar to the general case the random variables after the nonlinearity in polar coordinates is given by:

$$\begin{aligned} x_A &= g_A(s_A, \boldsymbol{\theta}) \\ x_\phi &= s_\phi + g_\phi(x_A, \boldsymbol{\theta}) \end{aligned} \tag{3.44}$$

Similarly, the inverse required for the transformation are given by:

$$\begin{aligned} s_A &= g_A^{-1}(x_A, \boldsymbol{\theta}) \\ s_\phi &= x_\phi - g_\phi(x_A, \boldsymbol{\theta})\,. \end{aligned} \tag{3.45}$$

With the Jacobian determinant

$$J = g'_A(g_A^{-1}(x_A, \boldsymbol{\theta}), \boldsymbol{\theta}) \tag{3.46}$$

the amplitude and phase distribution of a time sample after the nonlinearity is given as:

$$p_{x_{A/\phi}}(x_A, x_\phi \mid \boldsymbol{\theta}) = \frac{p_{s_{A/\phi}}(g_A^{-1}(x_A, \boldsymbol{\theta}), x_\phi - g_\phi(g_A^{-1}(x_A, \boldsymbol{\theta}), \boldsymbol{\theta})}{g'_A(g_A^{-1}(x_A, \boldsymbol{\theta}), \boldsymbol{\theta})}\,. \tag{3.47}$$

This is very similar to the general result (3.25). Combining the previous equations into a single equation yields the following distribution after the nonlinearity:

$$\begin{aligned} p_{x_{A/\phi}}(x_A, x_\phi \mid \boldsymbol{\theta}) &= \frac{g_A^{-1}(x_A, \boldsymbol{\theta})}{g'_A(g_A^{-1}(x_A, \boldsymbol{\theta}), \boldsymbol{\theta})\pi\sigma_s^2} \exp\left(-\frac{|g_A^{-1}(x_A)\exp(x_\phi - g_\phi(g_A^{-1}(x_A, \boldsymbol{\theta}), \boldsymbol{\theta}))|^2}{\sigma_s^2}\right) \\ p_{x_A}(x_A \mid \boldsymbol{\theta}) &= \frac{g_A^{-1}(x_A, \boldsymbol{\theta})}{g'_A(g_A^{-1}(x_A, \boldsymbol{\theta}), \boldsymbol{\theta})} \frac{2}{\sigma_s^2} \exp\left(-\frac{(g_A^{-1}(x_A))^2}{\sigma_s^2}\right) . \end{aligned} \tag{3.48}$$

This is a very important result. It shows that after the nonlinearity the distribution is still circular symmetric and hence independent of the phase s_ϕ. Since the amplitude does not contain any information about the phase distortion characteristic, the AM/PM nonlinearity cannot be estimated from a single distorted sample. If one observation does not contain any information about the phase distortion, this also applies to multiple independent observations. Therefore, *with i.i.d. time domain samples, a phase distortion cannot be estimated using a blind MLE.* On the other hand, if an AM/PM nonlinearity is present it does not interfere at all with the estimation of an AM/AM characteristic using this technique.

If the AM/AM characteristic is not bijective, similarly to the general case the distribution is given by

$$p_{x_A}(x_A \mid \boldsymbol{\theta}) = \sum_{v=0}^{V_g-1} \frac{g_A^{(v)}(x_A, \boldsymbol{\theta})}{g'_A(g_A^{(v)}(x_A, \boldsymbol{\theta}), \boldsymbol{\theta})} \frac{2}{\sigma_s^2} \exp\left(-\frac{(g_A^{(v)}(x_A))^2}{\sigma_s^2}\right) , \tag{3.49}$$

where $g_A^{(v)}$ is the v-th root as before. The amount of summands equals the actual amount of roots of the amplitude characteristic. This approach solves the problem of an exponentially growing amount of summands in the general case making non-bijective amplitude characteristics viable.

Special care must be taken with the domain of the resulting PDF which is usually limited. The reason is that most AM/AM nonlinearities have a maximum output amplitude A_{max} and therefore the domain of the inverse is the interval $[0, A_{max}]$. For amplitudes larger than A_{max} the value of the PDF is zero.

Transformation to carthesian coordinates Since the convolution is usually defined in terms of carthesian coordinates it is reasonable to undo the previous transformation to polar coordinates now.

$$p_x(x \mid \boldsymbol{\theta}) = \frac{1}{2\pi|x|} p_{x_A}(|x| \mid \boldsymbol{\theta}) . \tag{3.50}$$

This uses the fact that $p_x(x \mid \boldsymbol{\theta})$ exhibits circular symmetry, yielding a simpler expression as compared to (3.27).

Distribution after additive noise The additive white Gaussian noise process is also i.i.d. and each sample is distributed according to:

$$\begin{aligned} n &\sim CN(0, \sigma_n^2) \\ p_n(n) &= \frac{1}{\pi\sigma_n^2} \exp\left(-\frac{nn^*}{\sigma_n^2}\right) . \end{aligned} \tag{3.51}$$

The distribution after adding the noise is then given by the convolution of the noise and signal PDFs:

$$p_y(y \mid \boldsymbol{\theta}) = p_n(n) * p_x(x \mid \boldsymbol{\theta}) = \iint\limits_{\mathbb{C}} p_n(y') p_x(y - y' \mid \boldsymbol{\theta}) \mathrm{d}y' . \tag{3.52}$$

Generally it is still often not possible to find a closed-form analytical solution for (3.52) but the convolution now happens over just two dimensions. Using Fast Fourier Transformation (FFT)-based techniques it is possible to compute the convolution with complexity of order $O(L^2 \log L)$ where L is the amount of samples per convolution. An even faster method that exploits the circular symmetry of all involved signals is shown in section 5.3.

Example: SSPA nonlinearity the SSPA nonlinearity shall serve as an example for the resulting distributions. It is defined according to (2.25) and its inverse and first derivative are given by

$$g_A^{-1}(A, A_{max}, p) = \frac{A_{max} A}{\left(A_{max}^{2p} - A^{2p}\right)^{\frac{1}{2p}}} \text{ and } g_A'(A, A_{max}, p) = \left(1 + (\frac{A}{A_{max}})^{2p}\right)^{-1-\frac{1}{2p}} . \tag{3.53}$$

Combining these with the previous derivation allows to straight-forwardly derive the PDF after the nonlinearity analytically which is a lengthy expression that is omitted here. The resulting PDF after the noise has been calculated numerically using the

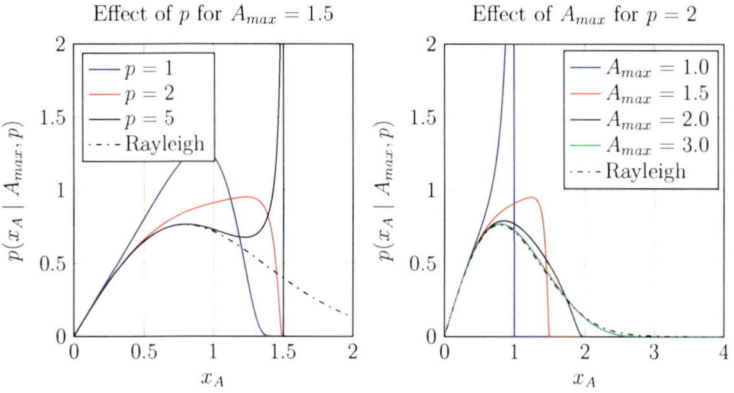

Figure 3.5.: Resulting SSPA densities before AWGN

FFT convolution approach with 1024 samples per dimension. Figure 3.5 shows some examples of the PDF $p_{x_A}(x_A \mid \boldsymbol{\theta})$ as defined in (3.48). All the density mass (i.e., area under the curve) larger than A_{max} is shifted to smaller values so that A_{max} marks the maximum output amplitude of the samples after the nonlinearity. When p is large and/or A_{max} is small, narrow and very high peaks close to A_{max} appear in the density. If further numerical processing of the density is to be done, care must be taken to sample appropriately to avoid aliasing effects. The numerical convolution of the densities becomes manageable under the given assumptions. Figure 3.6 shows the resulting distributions for multiple SNR values. Adding noise removes the limit A_{max} from the distribution and the density again has a nonzero value everywhere. For very low SNRs, the noise contribution starts to dominate and the density of the amplitudes approaches a Rayleigh distribution.

3.4.3. Maximum Likelihood Estimator

In the last section the density $p_y(y \mid \boldsymbol{\theta})$ of the observations conditioned on the unknown nonlinearity parameters has been derived. The task of the maximum likelihood estimator is now to find the parameters $\boldsymbol{\theta}_{opt}$ that maximize this function for a given set of observations. For N_o independent observations the problem

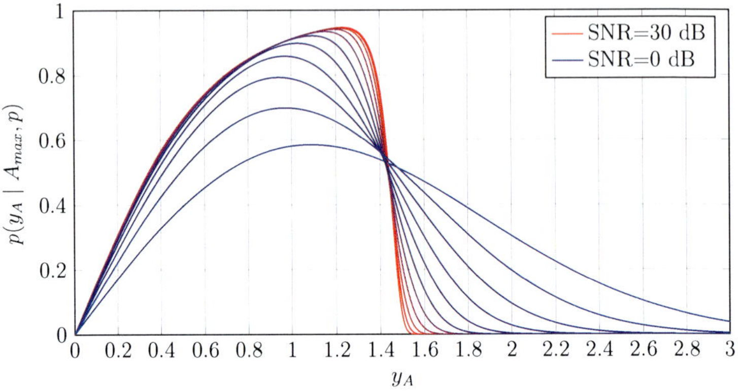

Figure 3.6.: Resulting densities for $A_{max} = 1.5$ and $p = 2$ after adding noise with 3 dB SNR increments

becomes:

$$\boldsymbol{\theta}_{opt} = \arg\max_{\boldsymbol{\theta}} \prod_{k=0}^{N_o-1} p_y(y_k \mid \boldsymbol{\theta}) = \arg\max_{\boldsymbol{\theta}} \sum_{k=0}^{N_o-1} \log p_y(y_k \mid \boldsymbol{\theta}) . \tag{3.54}$$

As mentioned earlier, the density is usually not available analytically so that the optimization has to be performed in a numerical fashion. Due to the monotonicity of the logarithm, optimization is performed over the sum of logarithms of each individual observation's density value. This is favorable since multiplying the density values of potentially thousands of observations can quickly lead to numerical instabilities with limited-precision floating or fixed point arithmetic. On the other hand, the logarithm compresses very large and very small numbers to a smaller range and summing those values is usually numerically stable. Figure 3.7 shows the resulting function that needs to be optimized for a range of candidate p and A_{max}. The true values and hence the desired estimates are marked with a white cross. The SNR ρ is defined as:

$$\rho = 10 \cdot \log_{10} \frac{\sigma_s^2}{\sigma_n^2} = -10 \cdot \log_{10} \sigma_n^2 , \tag{3.55}$$

since $\sigma_s^2 = 1$. In both cases, the function exhibits a well defined optimum close to the true values. In both cases a band can be seen where the likelihood is high. Nonlinearities with parameters in this range produce a very similar result when compared to a nonlinearity with the true parameters. As expected, the peak around

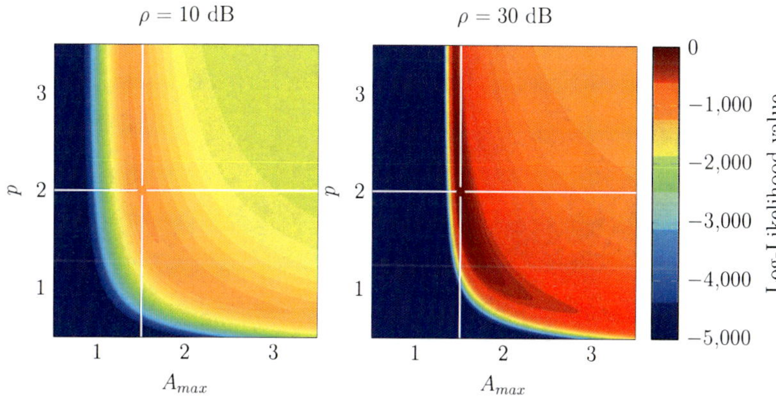

Figure 3.7.: Resulting log-likelihood for 10240 observations and a SSPA nonlinearity with $A_{max} = 1.5$, $p = 2$ and $\sigma_s^2 = 1$, true values marked white, log-likelihood normalized to zero at maximum of both plots

the optimum is lower and broader for the low-SNR case.

For the optimization algorithm the author obtained good results with the downhill simplex or Neader-Mead method [LRWW98] as implemented by Matlab's *'fminsearch'*-function. It is a general multivariate nonlinear optimization method that does not require derivatives. While it does not guarantee to converge to the global optimum, the author has very good experiences with this widely used and proven algorithm. The method requires an initial guess to start the optimization search. For many nonlinearities, only certain parameter ranges are useful. For the SSPA nonlinearity, the value of A_{max} is lower bounded by the minimum backoff that could be used and upper bounded by the maximum expected amplitude in the OFDM time domain samples. Similarly for $p > 5$, the resulting shape of the nonlinearity does not change significantly anymore. Therefore, randomly selecting some points in this range and using the best as an initial guess usually yields good results.

Before showing the results, a theoretical lower bound is derived for benchmarking the algorithm performance.

3.4.4. Cramer-Rao Lower Bound

The Cramer-Rao Lower Bound defines a lower limit of the achievable variance for unbiased estimators. It is defined as follows [Kay93b]

$$\operatorname{Var}\{\theta_i\} = \left[\underline{\boldsymbol{I}}^{-1}(\boldsymbol{\theta})\right]_{ii}, \tag{3.56}$$

where $\underline{\boldsymbol{I}}(\boldsymbol{\theta})$ is the Fisher information matrix which is defined as

$$[\underline{\boldsymbol{I}}(\boldsymbol{\theta})]_{ij} = -\operatorname{E}\left\{\frac{\partial^2 \ln p_{\boldsymbol{y}}(\boldsymbol{y} \mid \boldsymbol{\theta})}{\partial\theta_i \partial\theta_j}\right\}, \tag{3.57}$$

where the expectation is taken with respect to $\boldsymbol{y}$. If the elements of $\boldsymbol{y}$ are i.i.d., (3.57) is given by:

$$[\underline{\boldsymbol{I}}(\boldsymbol{\theta})]_{ij} = -N_o \cdot \operatorname{E}\left\{\frac{\partial^2 \ln p_y(y \mid \boldsymbol{\theta})}{\partial\theta_i \partial\theta_j}\right\}. \tag{3.58}$$

The Fisher information matrix can be understood as the average change of the density for a small variation in a parameter. The more sensitive the density reacts to parameter changes, the better it can be estimated. On the other hand, if the density does not change much when changing a parameter, estimators will usually require more observations for estimates with low variance. The bound is very practical to determine whether or not an estimator can still be improved.

Since $p_y(y \mid \boldsymbol{\theta})$ is only known numerically, applying (3.58) directly results in a double numeric differentiation followed by a numeric integration. While this is possible, it is computationally intensive and might provoke numerical errors. Firstly, by using the identity

$$\frac{\partial^2 \ln f(x,y)}{\partial x \partial y} = \frac{\frac{\partial^2}{\partial x \partial y} f(x,y)}{f(x,y)} - \frac{\frac{\partial}{\partial x} f(x,y) \cdot \frac{\partial}{\partial y} f(x,y)}{f^2(x,y)} \tag{3.59}$$

the natural logarithm can be replaced and (3.58) is reformulated as:

$$[\underline{\boldsymbol{I}}(\boldsymbol{\theta})]_{ij} = -N_o \cdot \operatorname{E}\left\{\frac{\frac{\partial^2}{\partial\theta_i \partial\theta_j} p_y(y \mid \boldsymbol{\theta})}{p_y(y \mid \boldsymbol{\theta})} - \frac{\frac{\partial}{\partial\theta_i} p_y(y \mid \boldsymbol{\theta}) \cdot \frac{\partial}{\partial\theta_j} p_y(y \mid \boldsymbol{\theta})}{p_y^2(y \mid \boldsymbol{\theta})}\right\}. \tag{3.60}$$

Now, using

$$\frac{\partial}{\partial\theta_i} p_y(y \mid \boldsymbol{\theta}) = \frac{\partial}{\partial\theta_i}\left[p_x(x \mid \boldsymbol{\theta}) * p_n(n)\right] = \left[\frac{\partial}{\partial\theta_i} p_x(x \mid \boldsymbol{\theta})\right] * p_n(n) \tag{3.61}$$

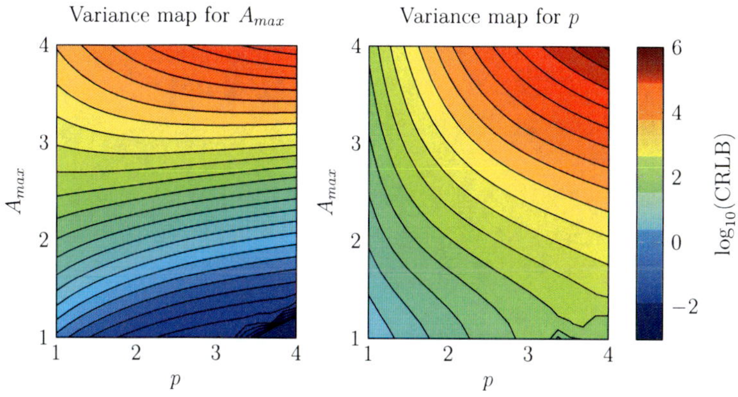

Figure 3.8.: CRLB maps for SSPA nonlinearity with $N_o = 1024$, $\rho = 10$ dB and $\sigma_s^2 = 1$

the partial derivatives can be performed analytically before convolving. While this increases the amount of convolutions, it is still favorable from a numerical point of view for reasons mentioned earlier. It should be noted that the derivatives are expressions of formidable length especially for the SSPA nonlinearity and hence omitted here. However, since algorithmic symbolic differentiation is a largely solved problem, the derivatives can be easily obtained using standard mathematics software.

Results The accuracy of the estimator depends on the nonlinearity parameters. Figure 3.8 shows the resulting CRLB-map for the estimation of a SSPA nonlinearity. The SNR ρ is defined as in (3.55). It can be seen that for large values of A_{max} and p, the estimation becomes inexact. This is reasonable since the amplifier is more linear and less samples will be subject to nonlinear distortion, hence the distribution changes less. It can also be seen that A_{max} is estimated with higher accuracy than p in almost any case. Those maps can be useful when designing the system, for example to obtain a good value for the required observation size.

Figure 3.9 shows a comparison between the CRLB and the empirical variance[4] of the maximum likelihood estimator. Most notable, with exception of the very low SNR region, *the estimator reaches the CRLB.* Therefore, the estimator gives the best possible unbiased performance in the high SNR region where the nonlinear effects

[4] Based on 2000 simulations per data point

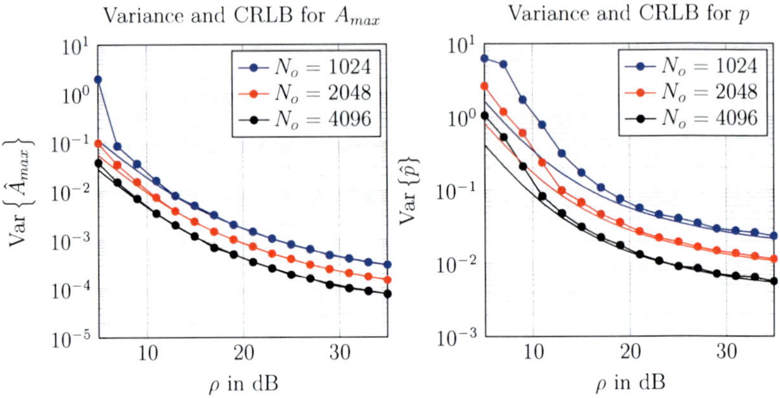

Figure 3.9.: Comparison of CRLB and empirical estimator variance for $p = 2$, $A_{max} = 1.5$ over different SNR values and observation sizes, Empirical curves with dotted markers

are dominating. By changing the amount of observations, it is possible to control the estimation accuracy almost arbitrarily. It should be noted that $N_o = 1024$ does not mean 1024 OFDM symbols but independent samples. For a system with 1024 subcarriers this is merely one single OFDM symbol. Therefore it can also be concluded that the estimator reaches a high accuracy with only very few observations.

Having an unbiased estimator that reaches the CRLB means that no significant further improvements are possible for this special case. Since the results are good the question arises if the method can be used in more realistic scenarios with oversampling and frequency selective channels. This is investigated now.

3.4.5. Application to oversampled OFDM and frequency selective fading

The main advantage of the special case under investigation is that each time sample is an independent observation and hence the required convolution becomes computationally manageable. The problems with (near-)singular covariance and convolution matrices are also not an issue. The goal is therefore to derive an algorithm that operates independently on each sample and still can produce good estimates when a guardband and frequency selective fading are present. Since these requirements are contradicting in a mathematical sense, the method will be suboptimal.

Handling guardband carriers As stated earlier, the covariance matrix of $\boldsymbol{s}$ is $\underline{\boldsymbol{C}}_{\boldsymbol{s}} = \sigma_s^2 \underline{\boldsymbol{\mathcal{I}}}_{N_c}$ if the elements of $\boldsymbol{s}$ are i.i.d. If that is not the case due to guardband carriers or per-subcarrier power allocation, the resulting covariance matrix is given by $\underline{\boldsymbol{C}}_{\boldsymbol{s}} = \sigma_s^2 \underline{\boldsymbol{\mathcal{I}}}_{N_c} + \underline{\boldsymbol{C}}_{\boldsymbol{s},c}$ where it is assumed that the average power per sample is still σ_s^2. The matrix $\underline{\boldsymbol{C}}_{\boldsymbol{s},c}$ describes the correlation between samples and $\mathrm{diag}(\underline{\boldsymbol{C}}_{\boldsymbol{s},c}) = \mathbf{0}$. When looking only at a single sample s_k, by marginalization it is found that $s_k \sim CN(0, \sigma_s^2)$.

So, each single sample alone is complex normally distributed as in the independent case. Only when looking jointly at multiple samples of the same symbol, the correlation gives some information about how the samples are dependent on each other. The idea is therefore to omit this information and assume the time samples as mutually independent, even though they are not.

Handling frequency selective filtering after the nonlinearity As shown in Figure 3.2, the received signal is given by

$$\boldsymbol{y} = \underbrace{\underline{\boldsymbol{H}} g(\boldsymbol{s}, \boldsymbol{\theta})}_{\boldsymbol{z}} + \boldsymbol{n} \,. \tag{3.62}$$

The resulting distribution of $\boldsymbol{z}$ is difficult to marginalize and hence looking at each sample independently is not easy. However, by applying zero-forcing equalization as follows

$$\boldsymbol{y}_{zf} = \underline{\boldsymbol{H}}^{-1} \boldsymbol{y} = g(\boldsymbol{s}, \boldsymbol{\theta}) + \underline{\boldsymbol{H}}^{-1} \boldsymbol{n} \tag{3.63}$$

the (assumed) independence between the distorted samples is restored and the noise is colored instead. The vector $\boldsymbol{H}$ describes the filtering in frequency domain so that:

$$\underline{\boldsymbol{H}} = \frac{\underline{\boldsymbol{W}} \, \mathrm{diag}(\boldsymbol{H}) \underline{\boldsymbol{W}}^H}{\sqrt{N_c}} \tag{3.64}$$

as described in section 3.3.3. Then,

$$\underline{\boldsymbol{H}}^{-1} = \frac{\underline{\boldsymbol{W}} \, \mathrm{diag}(\boldsymbol{H})^{-1} \underline{\boldsymbol{W}}^H}{\sqrt{N_c}} \tag{3.65}$$

and the resulting covariance matrix of $\boldsymbol{n}_{zf} = \underline{\boldsymbol{H}}^{-1} \boldsymbol{n}$ is given by $\underline{\boldsymbol{C}}_{\boldsymbol{n}_{zf}} = \sigma_n^2 \underline{\boldsymbol{H}}^{-1} \underline{\boldsymbol{H}}^{-1^H}$. After some matrix calculations it can be shown that each noise sample $n_{k;zf}$ after marginalization is distributed according to

$$n_{k;zf} \sim CN(0, \sigma_n^2 \cdot \overline{h}^2) \tag{3.66}$$

with

$$\overline{h}^2 = \frac{1}{N_c} \sum_{l=0}^{N_c-1} \left(\frac{1}{H_l} \right)^2 . \tag{3.67}$$

The algorithm After reception, the resulting samples are equalized in a zero-forcing manner to obtain $\boldsymbol{y}_{zf}$. The factor $\overline{h}^2$ is calculated as in (3.67). The samples $\boldsymbol{y}_{zf}$ are then used as basis for estimation as described earlier in this section. The noise power is set to $\sigma_s^2 \cdot \overline{h}^2$. Care should be taken that the noise power depends on the instantaneous channel realization and can change in one set of observations if it is large. Since observations with low instantaneous SNR do not improve the estimate, they should be omitted. The SNR-loss due to the frequency selective channel $\rho_{\overline{h}}$ is given by

$$\rho_{\overline{h}} = 10 \cdot \log_{10}(\overline{h}^2) \tag{3.68}$$

and all observations exceeding a certain SNR-loss threshold should be omitted. Since the nonlinearity is usually a slowly changing effect as compared to frequency selective fading, this is usually not a problem and will increase the quality of the estimates.

Estimation performance In Figure 3.10 the estimation variance is compared to the CRLB of a system with truly independent samples. Half of the system's subcarriers are guardband carriers which should be more than in most practical OFDM systems. The algorithm works very well in this case. While there is a visible performance loss as compared to Figure 3.9 it remains small over the whole SNR range and can easily be compensated by slightly increasing the amount of observations.

The picture looks different when a frequency selective channel is added as well. Figure 3.11 shows the resulting estimation performance. The channel exhibits an exponential Power Delay Profile (PDP) with 3dB decay per tap. The channel impulse response is assumed to be perfectly known and the algorithm has been configured to reject observations affected by a channel realization with more than 10dB of SNR-loss. Furthermore, all observations in one estimation are subject to the same channel impulse response. The shown curves can be seen as an ergodic estimation performance. As expected, there is a significant loss of estimation performance

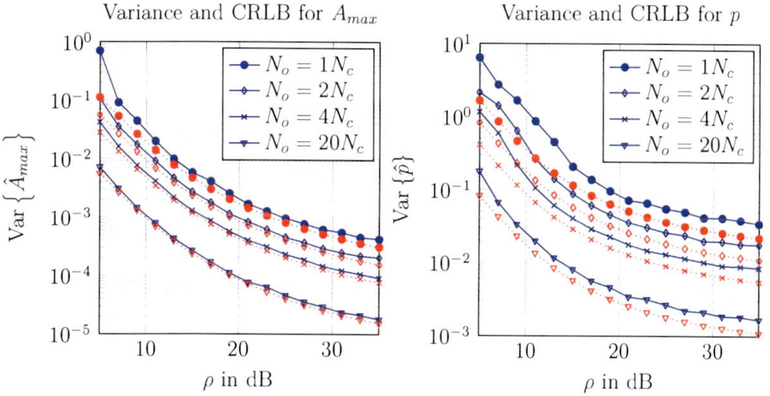

Figure 3.10.: CRLB and empirical variance for $p = 2$, $A_{max} = 1.5$, $N_c = 1024$, 50% guardband carriers over different SNR values and observation sizes (given in OFDM symbols), CRLB curves in red dotted style

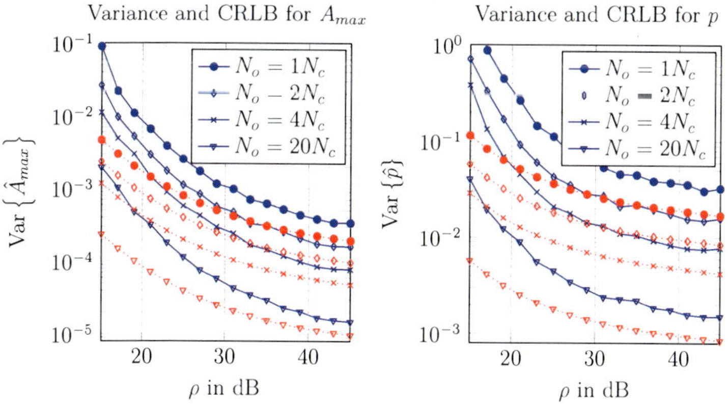

Figure 3.11.: CRLB and empirical variance after adding an exponential channel with 3dB decay per tap, other parameters as in Figure 3.10

caused by the SNR-loss due to the frequency selective fading. However, comparing the CRLB and empirical curves it can be seen that the SNR-loss is less than or roughly the same as the limit of acceptable SNR degradation. This implies that the curves can be improved by decreasing the acceptable SNR-loss at the cost of more rejected observations.

3.4.6. System Simulation Results

To judge if the quality of the estimates is sufficient, system simulations can compare the resulting bit error rates with the ones achieved under perfect parameter knowledge. The simulated system equals the one shown in Figure 2.11 but without a frequency selective channel to match the assumptions. The nonlinearity is of SSPA type. The system has $N_c = 1024$ subcarriers and the observation size is $N_o = 10$ OFDM symbols. The backoff γ is defined as the ratio between maximum output power and average signal power $\gamma = \frac{A_{max}^2}{\sigma_s^2}$. Figure 3.12 shows the results for Quaternary Phase Shift Keying (QPSK) and 16-QAM modulation.

The perfect curve is the achievable limit for the decision feedback detector in case of perfect nonlinearity knowledge and perfect feedback. The gap to the case without a nonlinearity is due to the decision feedback receiver. Since the nonlinear distortion $\boldsymbol{n_d}$ is assumed as noise and removed, in the best possible case this receiver will reach the undistorted result offset by a reduction of average output power. In the examples, the SNR gap is 1.4 dB for the case of $\gamma = 3$ dB and 4.1 dB for the case $\gamma = 0$ dB. For most cases, linear detection while ignoring the nonlinearity leads to an error floor. After one or two iterations the best possible result is reached with negligible SNR gap. Most importantly however, there is no visible gap between the case of perfect and estimated nonlinearity knowledge.

The estimator has also been implemented in a real demonstrator for tests with real overdriven analog amplifiers. See chapter 5 for details.

3.4.7. Conclusions

In this section, a low-complexity AM/AM estimator for an OFDM system with memoryless nonlinear distortions in the analog frontend has been proposed. The system is assumed to have no guardband carriers and no frequency selective filtering effects, e.g., multipath propagation. The CRLB for the problem has been derived and it was shown that the estimator reaches it very closely. It has been shown that

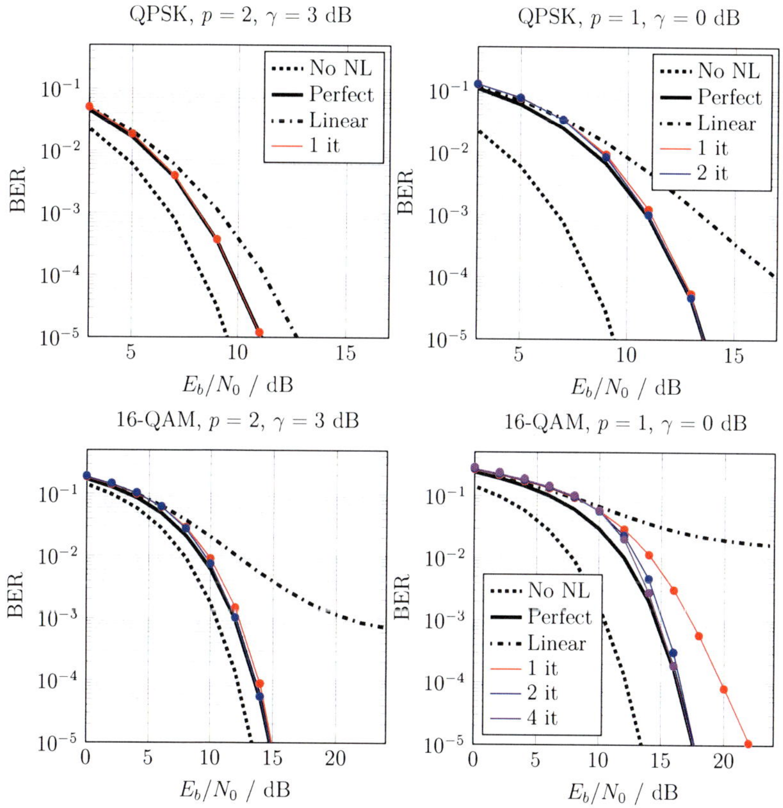

Figure 3.12.: Simulation results with different modulation schemes, comparing perfect (lines) and estimated knowledge (dots)

the performance gap between genie knowledge and estimation of the nonlinearity parameters becomes negligible for very small observation sizes such as 10 OFDM symbols.

Furthermore, a procedure has been presented that allows the application of the algorithm to systems with guardband and frequency selective fading. While this results in a loss of estimation performance, this loss stays manageable and can be mitigated by increasing the amount of observations.

The algorithm has been tested in system simulations and it was shown that there is no visible difference between the cases of perfect and estimated nonlinearity knowledge.

It was mathematically shown that in this simplified scenario, estimation of AM/PM nonlinearities is not possible. Low-complexity algorithms that allow this are investigated in the next section.

3.5. ML-Estimator for AM/PM nonlinearities in flat fading

3.5.1. Motivation

The method derived in the previous section solves the problem of the high complexity involved when numerically evaluating the likelihood of the estimation problem by assuming all received samples to be mutually independent. While this works well for AM/AM nonlinearities, it has been mathematically proven that AM/PM estimation is not possible using this method. So while AM/AM estimation can be done, even in the presence of an AM/PM nonlinearity, the goal of this section is to investigate low-complexity estimators for the AM/PM estimation problem [DF13].

3.5.2. Assumptions

It is assumed that there is non-zero correlation between the samples, i.e., the covariance matrix $\underline{C}_{\boldsymbol{s}}$ of the source vector $\boldsymbol{s}$ has non-zero off-diagonal elements. This is the regular case due to the presence of guardband carriers in most real OFDM systems. Furthermore it is assumed that there is either no AM/AM nonlinearity or its parameters are known, hence the vector $\boldsymbol{\theta}$ of parameters to estimate only refers to the parameters of the AM/PM nonlinearity. This assumption is valid since

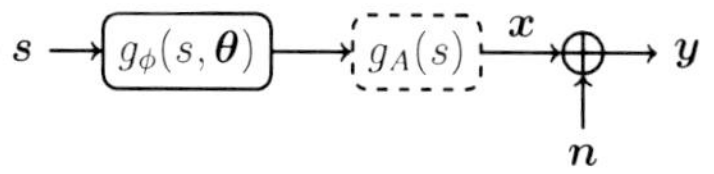

Figure 3.13.: Model for AM/PM estimation

the parameters of the AM/AM nonlinearity can and should be estimated using the method described in the last section. Lastly, flat fading is assumed although it is generally possible to apply the method described in section 3.4.5 when a frequency selective channel is present.

3.5.3. Derivation of the ML-Estimator

The estimation model is shown in Figure 3.13. The resulting probability density function is equal to the one derived earlier in section 3.3.3. It should be noted that all practical problems mentioned in section 3.3.4 apply. Especially the problem with a non-invertible AM/AM characteristic persists independently of whether the respective parameters are known or not.

AM/PM distortion only If there is no AM/AM distortion present in the system, a particularly compact density function is the result. This corner case is also practical to asses the principal performance that can be achieved. Hence, the focus is on this case in the following. By setting $g_A(\boldsymbol{s}_A) = 1$ it can be easily shown that the noise free received signal $\boldsymbol{x}$ is distributed according to

$$p_{\boldsymbol{x}}(\boldsymbol{x} \mid \boldsymbol{\theta}) = \frac{1}{\pi^{N_c} |\underline{\boldsymbol{C}}_{\boldsymbol{s}}|} \exp\left([\boldsymbol{x}^H \exp(j g_\phi(\boldsymbol{x}, \boldsymbol{\theta}))] \cdot \underline{\boldsymbol{C}}_{\boldsymbol{s}}^{-1} \cdot [\boldsymbol{x} \exp(-j g_\phi(\boldsymbol{x}, \boldsymbol{\theta}))] \right) , \tag{3.69}$$

where the phase rotations inside the brackets happen element-wise. The received signal is again the convolution of the distributions of $\boldsymbol{x}$ and $\boldsymbol{n}$:

$$p_{\boldsymbol{y}}(\boldsymbol{y} \mid \boldsymbol{\theta}) = p_{\boldsymbol{x}}(\boldsymbol{x} \mid \boldsymbol{\theta}) * p_{\boldsymbol{n}}(\boldsymbol{n}) . \tag{3.70}$$

Resulting distribution The resulting distribution functions are usually exhibiting too many dimensions to actually visualize. However, if the system only has two subcarriers, by fixing one subcarrier at a specific value, the conditional density can be visualized. The result is shown in Figure 3.14 for a system with correlation $[\underline{\boldsymbol{C}}_{\boldsymbol{s}}]_{0,1} = 0.8$ between the samples and an AM/PM nonlinearity according to the

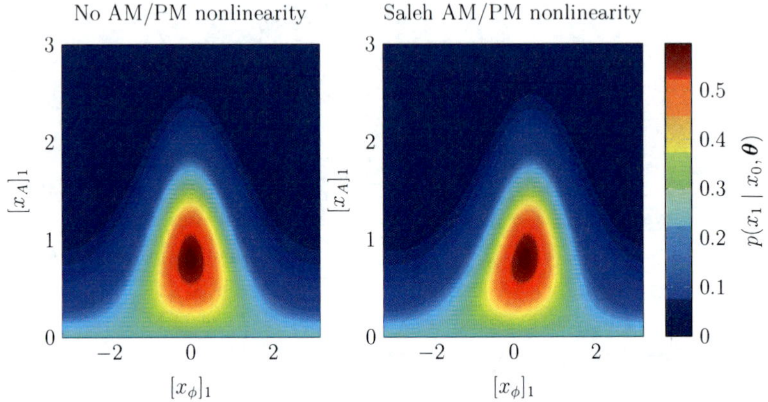

Figure 3.14.: Qualitative plot of $p_{\boldsymbol{x}}(x_1 \mid x_0 = 1, \boldsymbol{\theta})$ with and without TWTA nonlinearity

Saleh model with $\alpha_\phi = 1$ and $\beta_\phi = 2$. The first sample is fixed to be $x_0 = 1$. It can be seen that the conditional distribution $p_{\boldsymbol{x}}(x_1 \mid x_0, \boldsymbol{\theta})$ differs when an AM/PM nonlinearity is present. This indicates that a maximum likelihood estimator can be utilized to estimate $\boldsymbol{\theta}$.

However, even with just AM/PM distortion present, evaluating (3.70) numerically remains a quasi-impossible task, especially with real-time requirements. Therefore, multiple low-complexity implementations are presented now.

3.5.4. Low-complexity algorithms

Three algorithms are proposed which allow the approximate evaluation of $p_{\boldsymbol{y}}(\boldsymbol{y} \mid \boldsymbol{\theta})$.

Ignoring noise Since the density of $\boldsymbol{x}$ is known analytically and computationally complex calculations are only required to calculate the PDF of the noisy received signal $\boldsymbol{y}$, the easiest way to circumvent these computations is to assume a noise-free transmission. Especially in the high SNR-region, $\boldsymbol{y} = \boldsymbol{x} + \boldsymbol{n} \approx \boldsymbol{x}$ so that:

$$p_{\boldsymbol{y}}(\boldsymbol{y} \mid \boldsymbol{\theta}) = p_{\boldsymbol{x}}(\boldsymbol{y} \mid \boldsymbol{\theta})\,. \tag{3.71}$$

Large estimation errors are expected in the low SNR-region. However, if the signal is only mildly nonlinearly distorted, the noise will dominate over the effect of the nonlinearity in the low-SNR region. In those cases, not much can be gained by nonlinearity mitigation. Only in the high SNR-region the nonlinearity starts to become the dominating source of transmission errors.

While this method solves the problem of the computationally intensive numerical convolution, the problems with the (near-)singular covariance matrix and non-bijective AM/AM characteristics remain. Furthermore, large matrix multiplications are required, yielding a computational complexity of order $O(N_c^2)$ per evaluation.

Two-subcarrier decomposition For the AM/AM nonlinearity, the problems of a highly complex convolution, (near-)singular covariance matrix and non-bijective AM/AM characteristic were solved by assuming the source samples as mutually independent. The distribution after marginalization of all but one sample served as the distribution for this sample. Very good results have been achieved by this approach. Next to resulting in a much simpler convolution, the problem with (near) singular covariance matrices in the presence of guardband carriers also vanished.

Following the same line of thought, the smallest possible observation vector for AM/PM estimation must contain two elements since non-zero correlation between the samples is required to estimate the AM/PM nonlinearity using maximum likelihood estimators. So by selecting two elements i, j of $\boldsymbol{s}$, a new random vector $\boldsymbol{s}_{ij}$ can be built by means of marginalization:

$$\begin{aligned} \boldsymbol{s}_{ij} &= [\boldsymbol{s}_i, \boldsymbol{s}_j] \\ \boldsymbol{s}_{ij} &\sim CN(\mathbf{0}, \underline{\boldsymbol{C}}_{\boldsymbol{s}_{ij}}) \\ \underline{\boldsymbol{C}}_{\boldsymbol{s}_{ij}} &= \begin{bmatrix} \sigma_s^2 & [\underline{\boldsymbol{C}}_{\boldsymbol{s}}]_{ij} \\ [\underline{\boldsymbol{C}}_{\boldsymbol{s}}]_{ji} & \sigma_s^2 \end{bmatrix} \end{aligned} . \tag{3.72}$$

It has been shown earlier that estimation is not possible for independent samples. In the case of a normal distribution, the samples are independent if they are uncorrelated, i.e., $[\underline{\boldsymbol{C}}_{\boldsymbol{s}}]_{ij} = 0$. Similarly, estimation performance will be very poor when the covariance is small. Therefore, only samples exhibiting a very high covariance $|[\underline{\boldsymbol{C}}_{\boldsymbol{s}}]_{ij}|$ should be selected. Since the covariance matrix of the OFDM time vector is circulant, $N_c - 1$ two-subcarrier vectors exhibiting the same covariance can always be extracted from a single OFDM symol. The author had the best experiences using only sample-pairs with the highest covariance from one OFDM symbol. However, if the amount of observations is limited, sample pairs with lower correlation can be extracted as well.

Similarly to the AM/AM procedure, those two element vectors are now assumed as independent observations which dramatically reduces the dimensionality of the estimation problem. Compared to just ignoring the noise, decomposing the signal into independent two-element vectors reduces the computational complexity to $O(N_o)$ with N_o being the amount of two-element observations. Furthermore, the problems with (near-)singular covariance matrices are solved since $\underline{\boldsymbol{C}}_{\boldsymbol{s}_{ij}}$ will only be singular in the case of just one active subcarrier. Lastly, the problem of non-bijective AM/AM distortion is alleviated. If the characteristic has V_g roots, the sum in the PDF contains V_g^2 summands.

3.5.5. Two-subcarrier decomposition and convolution approximation

With just two samples per observation, numerically calculating the convolution of the distorted signal and noise distributions becomes a viable option. In principle, there are two options for performing the convolution.

Firstly, both $p_{\boldsymbol{x}_{ij}}(\boldsymbol{x}_{ij} \mid \boldsymbol{\theta})$ and $p_{\boldsymbol{n}_{ij}}(\boldsymbol{n}_{ij})$ can be sampled with L samples in each dimension. Using multidimensional FFT-techniques the calculations can be done with a complexity of order $O(L^4 \log L)$. The resulting convolved density is then interpolated to obtain the concrete densities for the observations $\boldsymbol{y}_{ij}$. One main limitation here is memory. Using standard 64-bit double-precision floating point numbers, each 4-dimensional array requires $8 \cdot L^4$ bytes of memory. For $L = 64$ this already means 128 MByte per array. Memory requirement increases to 2 GByte and 32 GByte per array for $L = 128$ and $L = 256$ respectively. Since multiple such arrays are required, practically $L = 64$ was the limit for the author. With such a low sampling resolution, aliasing effects can quickly occur. Especially for high SNR values, $p_{\boldsymbol{n}_{ij}}(\boldsymbol{n}_{ij})$ will rapidly decay and hence contain high frequencies as compared to $p_{\boldsymbol{x}_{ij}}(\boldsymbol{x}_{ij} \mid \boldsymbol{\theta})$. By requiring the same grid for both functions, it will rarely be sampled with more than three or five samples different from zero. In summary, the author was not able to get enough accuracy using this method for meaningful estimation.

To alleviate most of these problems, another type of calculation is possible. For a single observation $\boldsymbol{y}_{ij}$, the density function is given as

$$p_{\boldsymbol{y}_{ij}}(\boldsymbol{y}_{ij} \mid \boldsymbol{\theta}) = \iint_{\mathbb{C}^2} p_{\boldsymbol{x}_{ij}}(\boldsymbol{y}_{ij} - \boldsymbol{n}_{ij} \mid \boldsymbol{\theta}) \cdot p_{\boldsymbol{n}_{ij}}(\boldsymbol{n}_{ij}) \mathrm{d}\boldsymbol{n}_{ij} \,. \tag{3.73}$$

Instead of calculating (3.73) for a large grid of values and then interpolating those to obtain the density of the actual observations $\boldsymbol{y}_{ij}$, the convolution can be calculated just for the values of each individual observation and only the noise density needs to be sampled. Since the noise samples are independent, exhibit circular symmetry and are zero-mean, it is reasonable to sample it equally in each dimension and around the origin. Assuming L samples per dimension with equal distance d_q, let $\boldsymbol{q}$ be the vector whose elements contain the positions of the samples. The convolution of the discretized PDFs is then given as:

$$p_{\boldsymbol{y}_{ij}}(\boldsymbol{y}_{ij} \mid \boldsymbol{\theta}) \approx d_q^4 \cdot \sum_{\substack{\Re\{n_i\},\Im\{n_i\}\in\boldsymbol{q} \\ \Re\{n_j\},\Im\{n_j\}\in\boldsymbol{q}}} \cdots \sum p_{\boldsymbol{x}_{ij}}(\boldsymbol{y}_{ij} - \boldsymbol{n}_{ij} \mid \boldsymbol{\theta}) \cdot p_{\boldsymbol{n}_{ij}}(\boldsymbol{n}_{ij}) \,. \tag{3.74}$$

Since now only the noise density is sampled, $\boldsymbol{q}$ can be chosen so that aliasing and sampling values that are very close to zero are mostly avoided. The complexity of this method lies in the region $O(L^4 N_o)$ and hence depends on the amount of observations. The memory usage however only depends on the number of samples per dimension. Since this can usually be chosen much smaller than when sampling both $p_{\boldsymbol{x}_{ij}}$ and $p_{\boldsymbol{n}_{ij}}$ on the same grid, memory problems are largely avoided with this method.

3.5.6. Estimation performance

The following simulation setup was used to assess the estimation performance:

- 512 subcarrier OFDM, 256 guardband carriers
- QPSK and 16-QAM modulation
- AM/PM only TWTA nonlinearity with $\alpha_\phi = 2$ and $\beta_\phi = 2$
- $L = 3$ samples per dimension in convolution approximation
- 511 sample pairs with strongest correlation used per symbol

Figure 3.15 shows the results for the different approaches. The SNR ρ is defined as in (3.55). It shows the mean square error and its decomposition into estimatior variance and bias. Since all algorithms use inexact assumptions in order to handle the computational complexity, large biases in the low-SNR region are the result. This means that no matter how many observations are used, the squared bias will

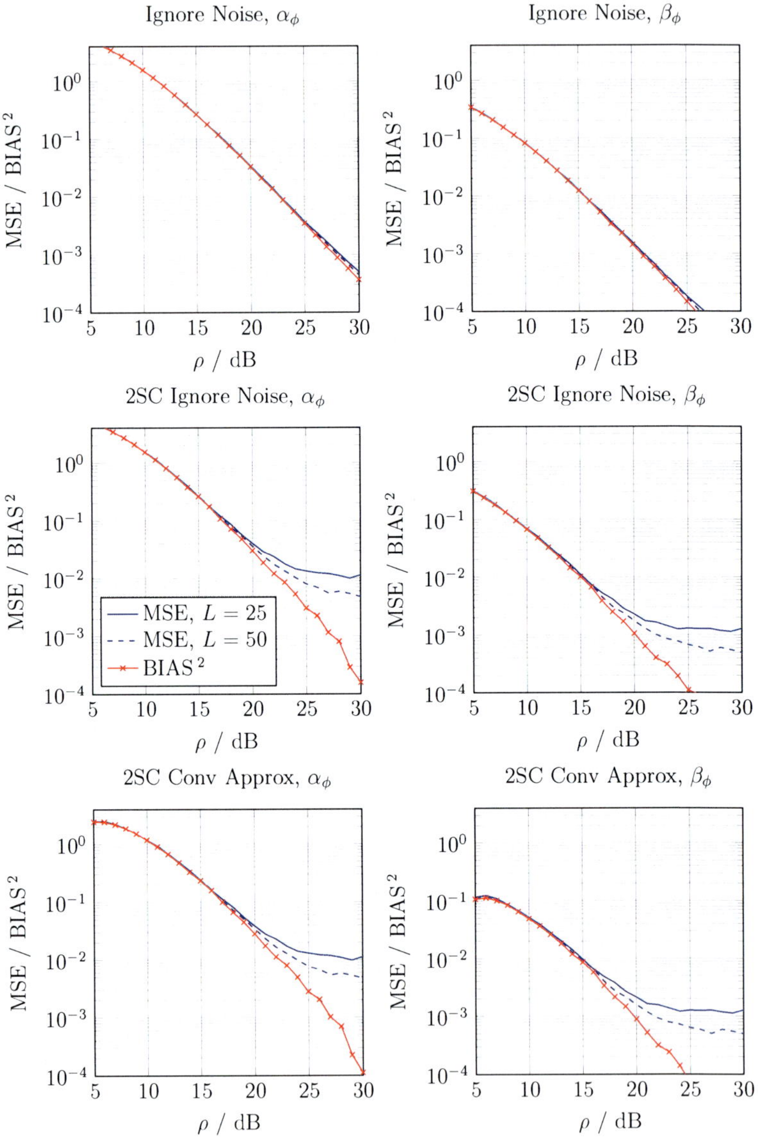

Figure 3.15.: Estimation performance of approximate AM/PM estimators

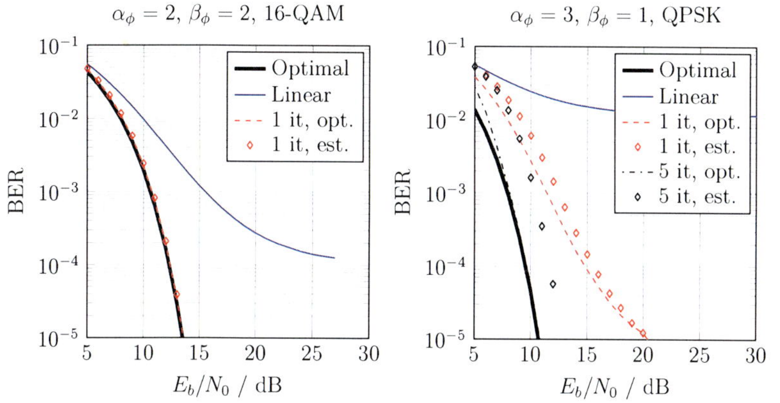

Figure 3.16.: System performance with decision feedback detector, comparison of optimal and estimated nonlinearity characteristic

always be a lower bound for the mean squared error. It can be seen that when simply ignoring the noise, the bias completely dominates the Mean Squared Error (MSE) over the whole SNR range. However, when using the two-subcarrier decomposition, the estimatior variance dominates starting at about 20dB SNR. The reason is that at higher SNR the samples with lower correlation can contribute information to the estimator which is ignored in the two subcarrier decomposition. In this range, doubling the amount of observation gives the expected halved estimator variance. For $L = 3$ samples per dimension, approximating the convolution gives only marginally better results in the very low SNR-region. However, the whole system performance is usually dominated by AWGN in those cases so that this method does not offer additional value towards ignoring the noise. Increasing L might improve the performance but quickly increases the required simulation complexity beyond the possibilities of the author and what is possible in many current real-time systems.

Figure 3.16 shows the resulting bit error rates for a system with decision feedback detection. It can be seen that despite gross misestimates the best possible performance can be reached for the less severe nonlinearity with $\alpha_\phi = \beta_\phi = 2$. However, there is a significant performance hit for the more severe nonlinearity with $\alpha_\phi = 3$, $\beta_\phi = 1$. In that case, a significant SNR gap of 2 dB at a bit error rate of 10^{-3} remains. However, it can also be seen that for higher SNR values, the gap between estimated and genie knowledge shrinks, inidicating better performance in the high SNR-region. There is no visible performance difference among the three methods. Since the method of ignoring the noise with decomposition into two-subcarrier subvectors has the lowest complexity, this method is recommended by the author if the AM/PM estimation is to be performed blindly.

3.6. Conclusions

In this chapter, feed forward methods for the estimation of AM/AM and AM/PM characteristics have been presented. This approach has the advantage of putting very low requirements on the receiver. For many cases, only the I/Q-samples right after reception are required. In case of a frequency selective channel, knowledge about this channel is required, as well, which can be assumed present in most OFDM receivers.

The estimation is done using maximum likelihood techniques and the estimator for the most common case has been mathematically derived. It has been shown that this estimator is not practically implementable due to a very high-dimensional convolution.

Firstly, a reduced-complexity estimator for AM/AM nonlinearities in frequency flat channels and without guardband carriers has been derived. It is based on the assumption, that all samples belonging to one OFDM symbol are i.i.d. It was shown that the estimator reaches the Cramer-Rao Lower Bound over a wide SNR-range. Two extensions have been proposed to employ this algorithm in case of oversampled OFDM signals and the presence of a frequency selective channel. While there is a performance loss, it can be controlled by rejecting low SNR observations and very good estimation performance is still possible for very small observation windows. It has been shown that AM/PM estimation is impossible when the samples are assumed to be mutually independent. The method allows the estimation of an AM/AM nonlinearity when an AM/PM nonlinearity is present.

Secondly, three reduced-complexity estimators for AM/PM nonlinearities in frequency flat channels have been presented. Since the samples must not be independent if AM/PM estimation is to be performed, the main issue is to limit the complexity of the required convolution. To that end, all algorithms operate under assumptions that introduce large estimation errors especially in the low-SNR region. Since the main source of error is bias in this region, increasing the amount of observations does not improve estimation performance. However, for less severe AM/PM nonlinearities, a very good estimation performance is possible that allows similar system performance as in the case of genie knowledge.

Especially in the presence of AM/PM nonlinearities, there are practical limits to real-time computability of the estimates. For neither AM/AM nor AM/PM estimation, the optimal ML estimator can be implemented due to very high complexity and special assumptions need to be made in order to get the computational to manageable levels. This however decreases the estimation performance significantly. In the next chapter feedback-based estimation methods are investigated that alleviate most of these problems.

4. Decision Feedback Estimation

4.1. Introduction

In the previous chapter, feed forward estimation methods have been investigated. Their advantage is low requirements on the transmitter but Amplitude to Phase (AM/PM) estimation is inaccurate or computationally very expensive. The nonlinear receiver employs a decision feedback technique to iteratively improve the quality of nonlinearity mitigation. Up to this point, the blind estimator is not using this iterative method.

The goal of this chapter is to further improve the estimator by harnessing the information of the feedback. This estimator, in its most general form, should be able to jointly estimate Amplitude to Amplitude (AM/AM) and AM/PM nonlinearities while still exhibiting a very low complexity. The overall estimation performance is expected to surpass all feed forward methods due to the additional information of the feedback [DF14].

4.2. System Model

The system model is shown in Figure 4.1. It is very similar to the one used in chapter 3, see Figure 3.1. It again consists of a decision feedback detector that gains its knowledge about the nonlinearity characteristic by means of blind estimation. The difference is that the estimator is not only getting the received samples $\boldsymbol{y}$ but also the estimates on the original signal $\hat{\boldsymbol{s}}$ before the nonlinearity and channel influences that are calculated in every iteration of the decision feedback detector by means of remodulation from the previously detected data. This gives the estimator a vast amount of additional information that cannot only be used to improve the estimation performance but also reduce the computational complexity.

Note that the system does not require a decision feedback architecture to be used. An estimate $\hat{\boldsymbol{s}}$ of the transmitted time samples can usually be obtained in

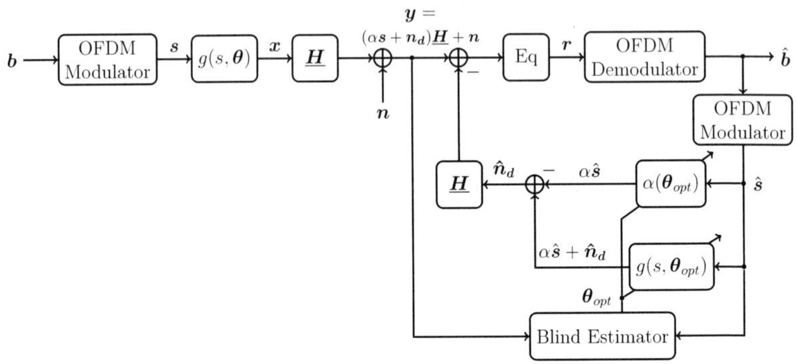

Figure 4.1.: OFDM transmission with feedback based blind estimator in receiver

any communications receiver that outputs the received information bits. Feedback, however, dramatically improves the estimation performance and hence, the method performs best in iterative architectures.

4.3. Blind Feedback Estimator

4.3.1. Assumptions

In the following derivation it is assumed that $\hat{\boldsymbol{s}} = \boldsymbol{s}$. It can later be seen that this assumption dramatically reduces the resulting computational complexity. The basic idea is that in each iteration, the error between $\hat{\boldsymbol{s}}$ and $\boldsymbol{s}$ is reduced until they become almost identical. In this case, the decision feedback detector offers near maximum likelihood performance. It was already seen in chapter 3 that performance improvements are possible despite large estimation errors in $\boldsymbol{\theta}$. This indicates that an iterative approach might be successful. The same assumption forms the basis of the very successful decision feedback detector.

Furthermore, the frequency selective channel $\underline{\boldsymbol{H}}$ and the variance of the Additive White Gaussian Noise (AWGN) σ_n^2 are assumed as perfectly known. This is mainly done to assess the estimation performance under otherwise perfect conditions. Results for a system with more practical assumptions are presented in chapter 5.

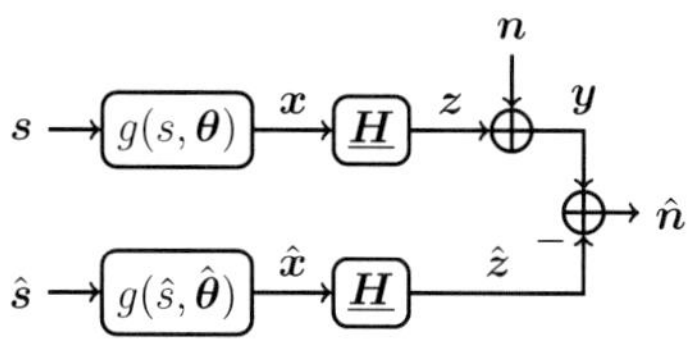

Figure 4.2.: Estimation model for Blind Feedback Estimator

4.3.2. Estimator derivation

The basic principle for the estimation is shown in Figure 4.2. The estimator gets the received values $\boldsymbol{y}$ and an estimate of the transmitted symbols $\hat{\boldsymbol{s}}$ as input. For every candidate solution $\hat{\boldsymbol{\theta}}$, the signal after the frequency selective channel is calculated as:

$$\hat{\boldsymbol{z}} = g(\hat{s}, \hat{\boldsymbol{\theta}})\underline{\boldsymbol{H}} = \hat{\alpha}\hat{\boldsymbol{s}}\underline{\boldsymbol{H}} + \hat{\boldsymbol{n}}_d\underline{\boldsymbol{H}}\,. \tag{4.1}$$

This signal is then subtracted from the received signal $\boldsymbol{y}$ to obtain estimates of the AWGN samples

$$\hat{\boldsymbol{n}}(\hat{\boldsymbol{s}}, \hat{\boldsymbol{\theta}}) = \boldsymbol{y} - \hat{\boldsymbol{z}} = \boldsymbol{n} + \underbrace{(\alpha\boldsymbol{s} - \hat{\alpha}\hat{\boldsymbol{s}} + \boldsymbol{n}_d - \hat{\boldsymbol{n}}_d)\underline{\boldsymbol{H}}}_{n_e}\,, \tag{4.2}$$

where $\boldsymbol{n}_e$ represents the combined error signal that results from the estimation errors in $\hat{\boldsymbol{s}}$ and $\hat{\boldsymbol{\theta}}$. Following the same line of thought as with the Bussgang theorem, $\boldsymbol{n}_e$ can be modelled as a random variable with a zero-mean distribution. Except for special cases, $\boldsymbol{n}_e$ will usually exhibit a non-Gaussian distribution with a density function that is difficult to find. However, there is an estimation method that does not require knowledge of this Probability Density Function (PDF).

Likelihood based estimation The density of the additive noise $\boldsymbol{n}$ is usually known and easy to evaluate. Following the same line of thought as for the maximum likelihood estimator, the following optimization

$$\boldsymbol{\theta}_{opt} = \arg\max_{\hat{\theta}} p_n(\hat{\boldsymbol{n}}) \tag{4.3}$$

leads to the parameter set having the highest likelihood. Note that this is not the same as the Maximum Likelihood Estimator (MLE). The MLE finds the conditional density that yields the highest likelihood for a fixed set of observations while this method finds the parameters that generate the noise estimate yielding the highest likelihood for a fixed noise distribution.

The method can be used for arbitrary noise distributions and hence is very general. It is very common to model $\boldsymbol{n}$ as additive white Gaussian noise. This results in an especially simple and practically very relevant estimation method.

Minimum variance estimation Under the assumption of additive white Gaussian noise, the distribution p_n is given as:

$$p_n(\hat{\boldsymbol{n}}) = \prod_{k=0}^{N_c-1} p_n(\hat{n}_k) = \prod_{k=0}^{N_c-1} \frac{1}{\pi\sigma_n^2} \mathrm{e}^{-\frac{|\hat{n}_k|^2}{\sigma_n^2}} . \tag{4.4}$$

Since optimization can equivalently happen on the logarithm, this can be further rewritten as:

$$\log p_n(\hat{\boldsymbol{n}}) = -N_c \log(\pi\sigma_n^2) - \frac{1}{\sigma_n^2} \sum_{k=0}^{N_c-1} |\hat{n}_k|^2 . \tag{4.5}$$

N_c and σ_n^2 are independent of the candidate $\boldsymbol{\theta}$ and have no significance in the maximization so that only the sum remains. Hence, (4.3) can be rewritten as:

$$\boldsymbol{\theta}_{opt} = \arg\min_{\hat{\boldsymbol{\theta}}} \sum_{k=0}^{N_c-1} |\hat{n}_k|^2 . \tag{4.6}$$

(4.6) is equivalent to finding the minimum of the sample variance of the noise estimates derived from the observations. This simplification is very handy since it is independent of σ_n^2 and hence, the noise variance does not need to be known.

Caveats It has been shown that the estimated parameters will on average be correct when $\hat{\boldsymbol{s}} = \boldsymbol{s}$, i.e., $\mathrm{E}\{\boldsymbol{\theta}_{opt}\} = \boldsymbol{\theta}$. However, this is generally not the case if $\hat{\boldsymbol{s}} \neq \boldsymbol{s}$. Instead, the best estimates will usually exhibit a bias that tends to grow the larger the bigger the deviation between estimated and actual feedback is. However, deriving an analytical relationship is rather difficult and also strongly depending on the other system components.

Advantages The main advantage of iterative estimation is the simplicity of the method especially if the common case of i.i.d. noise samples can be assumed. For each candidate $\hat{\boldsymbol{\theta}}$, only the nonlinearity needs to be applied, which is possible in $O(N_c)$ for memoryless nonlinearities and, if present, the effect of the frequency

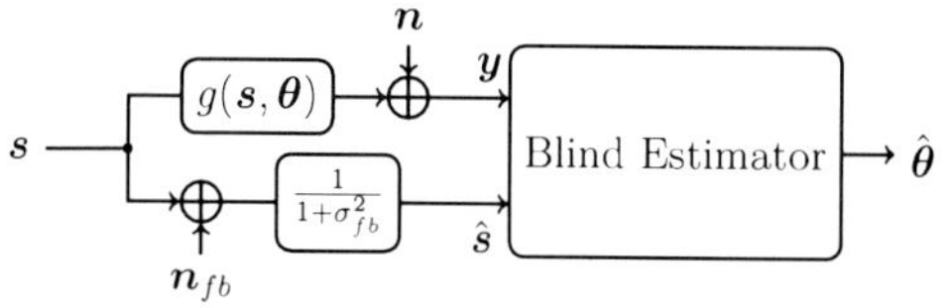

Figure 4.3.: Test fixture to evaluate principal iterative estimation performance

selective channel $\underline{\boldsymbol{H}}$ has to be calculated which can usually be done in the frequency domain with order $O(N_c \log N_c)$. Calculating (4.6) is a $O(N_c)$ operation yielding a total complexity of order $O(N_c \log N_c)$ which enables the usage in real-time systems. Calculating the optimization for different noise distributions might be more involved if the evaluation of the noise density is computationally complex.

Furthermore, there is no requirement that the transmitted signal $\boldsymbol{s}$ exhibits a Gaussian distribution anymore. Hence, this method could potentially be applied to other modulation-schemes as well. However, further research is required to verify this. Similarly, no more constraints are placed on the shape of the nonlinearity. Especially, there is no more complexity issues if the nonlinearity exhibits a non-monotonous or discontinuous PDF. Finally, the noise can also exhibit an arbitrary distribution. However, for efficient calculation, the noise density should be easy to evaluate. The generality paired with a very low complexity renders this a very powerful framework for nonlinearity estimation. Hence, the main goal for this estimator, as stated in the introduction of this chapter, is achieved.

4.3.3. Results

Since a fully simulated transmission system has too many influences, the principal performance is evaluated firstly using the scheme shown in Figure 4.3. The signal $\boldsymbol{s}$ is generated as white Gaussian noise with unit variance (i.e., $\sigma_s^2 = 1$) in accordance with the Gaussian assumption for OFDM signals, nonlinearly distorted with a Saleh nonlinearity and superimposed by AWGN to produce the received signal $\boldsymbol{y}$. The Signal to Noise Ratio (SNR) ρ between the information signal $\boldsymbol{s}$ and the noise $\boldsymbol{n}$ is then defined as:

$$\rho = 10 \cdot \log_{10} \frac{\sigma_s^2}{\sigma_n^2} \text{ dB} . \tag{4.7}$$

The feedback signal is generated by superimposing the original signal $\boldsymbol{s}$ with AWGN $\boldsymbol{n}_{fb}$ of variance σ_{fb}^2 and afterwards renormalizing the total power to one in order to

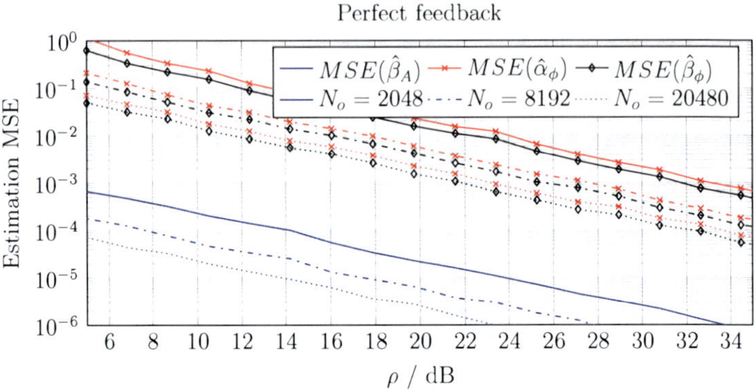

Figure 4.4.: Iterative estimation performance for $\beta_A = 0.3$, $\alpha_\phi = \beta_\phi = 2$ and perfect feedback $\hat{\boldsymbol{s}} = \boldsymbol{s}$, α_A is fixed to one and not subject to estimation

generate a feedback signal $\hat{\boldsymbol{s}}$ that resembles the result of a remodulation with errors. The feedback SNR ρ_{fb} is then defined as:

$$\rho_{fb} = 10 \cdot \log_{10} \frac{\sigma_s^2}{\sigma_{fb}^2} \text{ dB}\,. \tag{4.8}$$

The nonlinearity is then jointly estimated by the blind estimator.

Figure 4.4 shows the resulting estimation Mean Squared Error (MSE) for the case of perfect feedback $\hat{\boldsymbol{s}} = \boldsymbol{s}$, i.e. $\sigma_{fb}^2 = 0$. Due to the larger influence of the AM/AM characteristic, the parameter β_A is estimated with much higher accuracy than the AM/PM parameters. Furthermore, there is no estimation bias in this case since the assumption of perfect feedback is met. Hence, doubling the amount of observations shows the expected reduction of the estimation variance by factor two. By increasing the amount of observations, the estimation accuracy can therefore be controlled to meet the system's requirements. This holds even in case of time-varying frequency selective channels since each observation can have a different channel estimate.

When the assumption of perfect feedback is not met, the estimation is subject to bias errors and hence an error floor of the MSE that cannot be lowered by using more observations. This is shown in Figure 4.5 for the example of the parameter β_A and a feedback SNR of 20 dB. It can be seen that the MSE is limited by the square of the bias creating an error floor.

Figure 4.6 shows the resulting estimation MSE for β_A and α_ϕ and different values

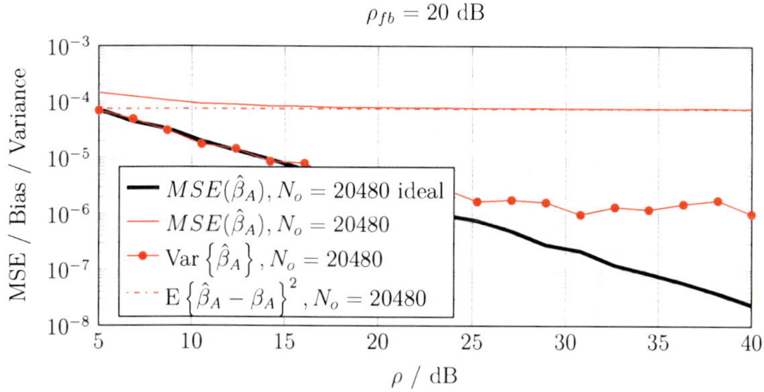

Figure 4.5.: Effects of imperfect feedback on iterative estimation, parameters as in Figure 4.4

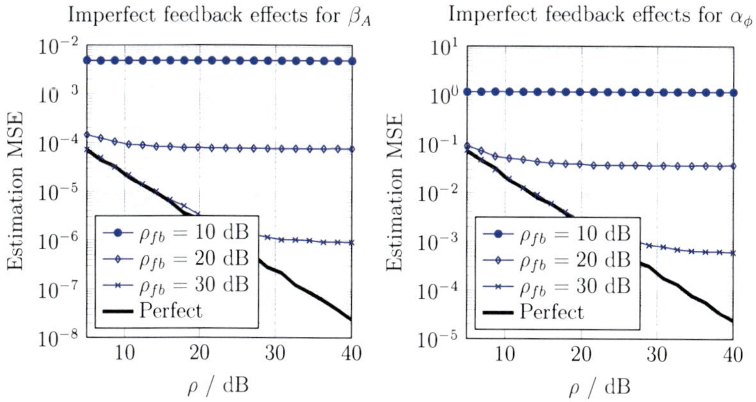

Figure 4.6.: Effects of imperfect feedback on the estimation for different feedback SNRs ρ_{fb}

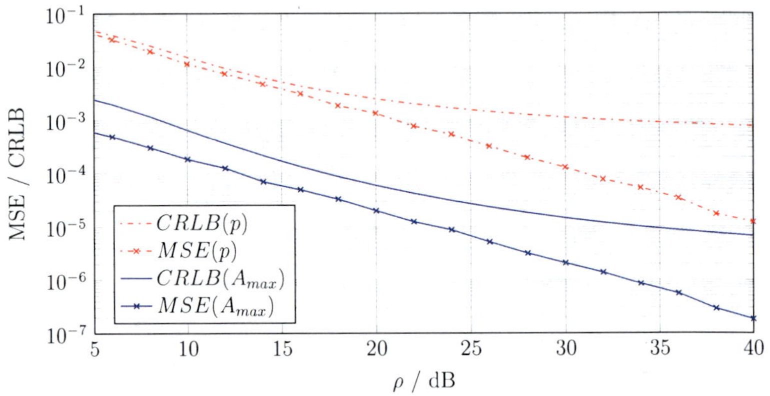

Figure 4.7.: Comparison of CRLB from section 3.4.4 with iterative estimation MSE (first iteration) for SSPA nonlinearity with $p = 2$ and $\gamma = 3$ dB

of the feedback SNR ρ_{fb}. It can be seen that the bias quickly dominates the total estimation error. However, it is also seen that the bias quickly vanishes when the feedback signal becomes better. This is a good indicator that an iterative approach can quickly converge to low estimation errors.

Finally it is very interesting to compare the resulting estimation MSE with the CRLB obtained for the maximum likelihood based estimator from section 3.4.4 where no feedback is used. This comparison is shown in Figure 4.7 for an AWGN channel. The iterative estimation performance is plotted for the first iteration and it can be seen that it outperforms the best possible performance of the non-iterative estimator over the whole SNR range. Furthermore, there is no improvement from further iterations in terms of estimation MSE. It has shown to be consistent over all modulation schemes that for an AWGN channel the first iteration already gives the best possible estimation performance. Further iterations may still improve the resulting bit error rate, though. It is the additional information contained in the feedback that allows to increase performance beyond the theoretical possibilities of the blind feed forward estimator. Therefore, if feedback is available, the iterative estimation method is superior.

System simulation results All system simulations were performed using the following parameters:

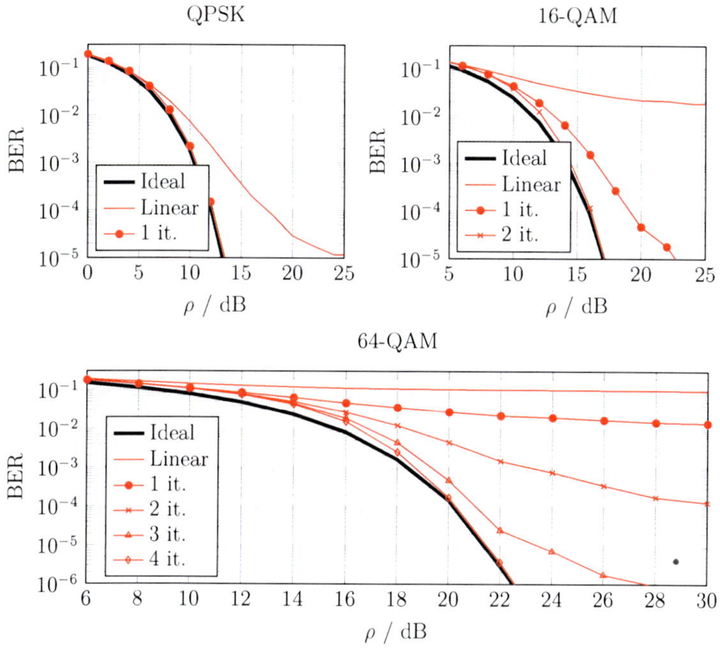

Figure 4.8.: Bit error rates for joint TWTA nonlinearity estimation with $\alpha_A = 0.3$, $\alpha_\phi = \beta_\phi = 2$ and an AWGN channel

- 512 subcarrier OFDM, 256 guardband carriers
- Quaternary Phase Shift Keying (QPSK), 16-QAM and 64-QAM modulation
- SSPA or Travelling Wave Tube Amplifier (TWTA) nonlinearities
- 40 symbols observation size (20480 samples)
- AWGN channel

The first system simulation consists of a jointly estimated TWTA nonlinearity with parameters $\beta_A = 0.3$, $\alpha_\phi = \beta_\phi = 2$. The parameter $\alpha_A = 1$ was fixed since it is indistinguishable from a channel attenuation. The results are shown in Figure 4.8. It can be seen that for linear reception all modulation schemes produce an error floor due to the additive distortion term introduced by the nonlinearity. Furthermore,

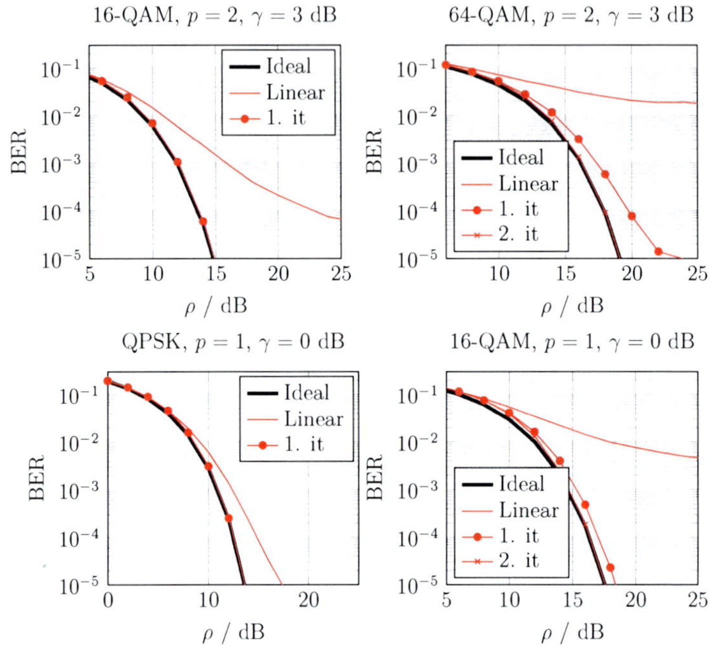

Figure 4.9.: Bit error rates for SSPA nonlinearity estimation in an AWGN scenario

after just four iterations, even for 64-QAM, the best possible performance is reached in the high SNR region. For QPSK and 16-QAM only one respectively two iterations are required for this. There is a small SNR loss in the low to medium SNR region but for the often important boundary of $BER = 10^-3$ the SNR loss is less than 0.3 dB even for 64-QAM modulation.

For comparison with the earlier results, Figure 4.9 shows the resulting Bit Error Rate (BER) curves for a jointly estimated SSPA nonlinearity with different parameters. As expected from the earlier results, in all cases the best possible performance is reached in at most two iterations.

4.4. Decision Feedback Estimation under Imperfect Channel Knowledge

4.4.1. Motivation

The decision feedback estimation methods discussed in this chapter promise better performance at lower computational complexity as compared to feed forward estimation methods described in chapter 3. However, due to the dependency on feedback, the estimation performance will always be tightly coupled to many other receiver components. One particularly important component here is the estimation of the frequency selective channel since it is used not only in the nonlinearity estimation, but also in the mitigation part of the decision feedback receiver. Perfect knowledge of the frequency selective channel was always assumed up to this point. In the following, this assumption is lifted and a channel estimation algorithm is used that will produce imperfect results.

The goal is therefore to test the iterative feedback estimator in a scenario with imperfect channel knowledge resembling a less ideal system. It also serves as the first step towards an actual implementation since there is no unrealistic perfect knowledge anymore. Hence, with the addition of a suitable synchronization, this simulation can be readily used in a hardware in the loop scenario.

4.4.2. System model

Figure 4.10 shows the block diagram of the whole simulation chain. Most notable difference to the former simulations is that the channel is no longer known perfectly but has to be estimated. To that end, pilot subcarriers are inserted into the datastream which are modulated with pseudo-random QPSK symbols. The channel estimator works by linearly interpolating between the pilot symbols separately for amplitude and phase. From the structure it follows that a faulty channel estimation influences both, estimation and mitigation. It is therefore especially interesting to observe the effect of imperfect channel knowledge.

The simulation still assumes perfect synchronization and no hardware impairments except for the nonlinear distortions. The channel is applied by means of multiplication with the circular convolution matrix $\underline{\boldsymbol{H}}$ which equals application in the frequency domain and allows for the omission of the cyclic prefix which hence is implicitly assumed to be longer than the channel delay spread so that no Inter Symbol

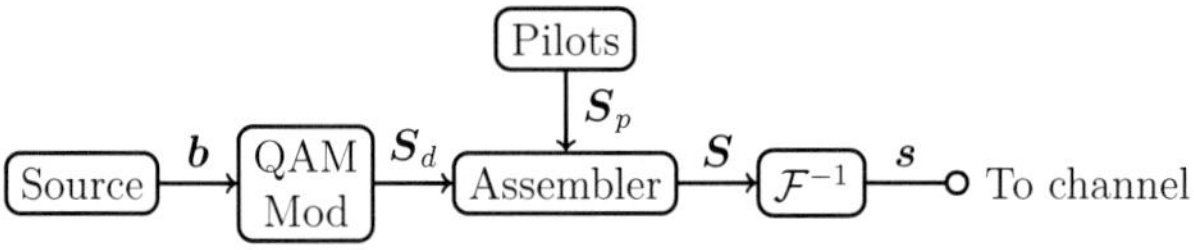

(a) Transmitter

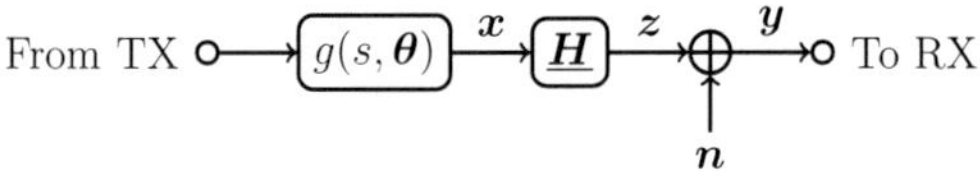

(b) Channel model

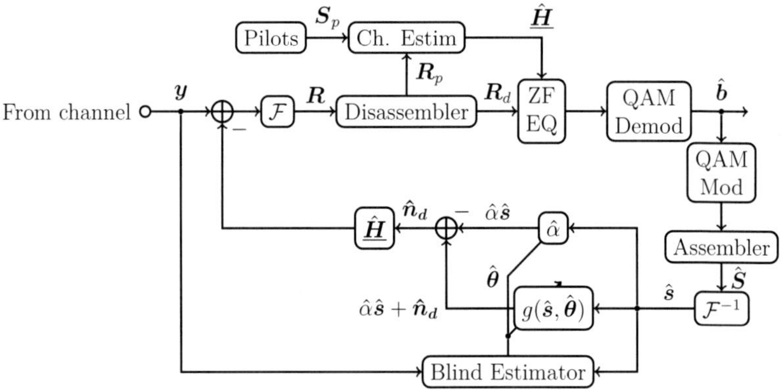

(c) Receiver

Figure 4.10.: Block diagram of the simulation chain for the iterative estimator

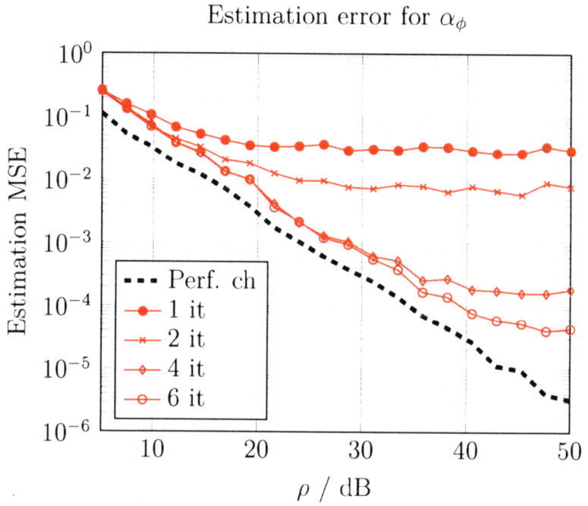

Figure 4.11.: Estimation performance with perfect and imperfect channel knowledge for 16-QAM transmission and a TWTA nonlinearity with $\beta_A = 0.3, \alpha_\phi = \beta_\phi = 2$

Interference (ISI) occurs.

4.4.3. Results

The system has been parameterized to have 512 subcarriers in total, 256 of which are guardband carriers. The pilots are spaced four subcarriers apart, resulting in 64 pilot carriers and 192 data carriers. The channel is modelled as a tapped delay line. Each tap has a Rayleigh distribution of the amplitude and a uniform phase distribution. The power delay profile is exponential with 3 dB decay per tap. Data is processed in blocks of 40 OFDM symbols each. The whole block is used as observations for the estimator. Each OFDM symbol in one block is affected by the same channel realization and channel realizations between blocks are independent of one another. Information about previous nonlinearity estimates $\hat{\boldsymbol{\theta}}$ is reused between iterations of one block but each block is estimated completely separate of one another.

Estimation performance Figure 4.11 shows a comparison of the estimation performance with and without perfect channel knowledge. In earlier simulations using perfect channel knowledge, no significant gains in estimation performance have shown after the first iteration. However, Figure 4.11 shows that with imperfect channel

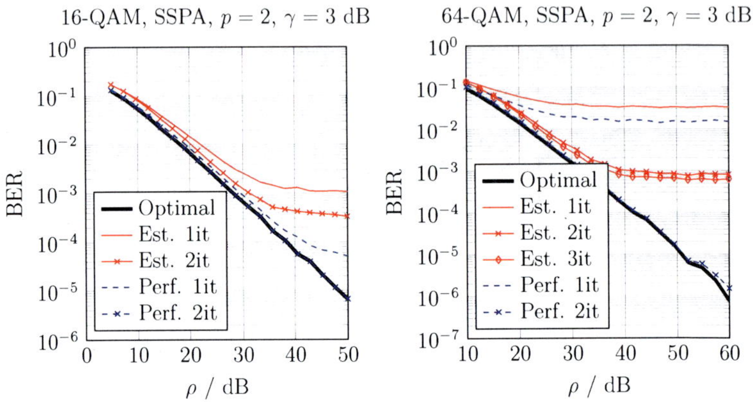

Figure 4.12.: BER comparison for SSPA nonlinearity with perfect and estimated channel knowledge, $p = 2, \gamma = 3$ dB

knowledge, the estimation performance improves over several iterations. This is because the removal of nonlinear distortion allows for better channel estimates. Since they are used in the blind estimation process, in turn, better channel estimates improve the estimation performance. This behavior has shown consistently for all simulations.

System performance Figure 4.12 shows the resulting BERs for a SSPA-type nonlinearity and compares the case of perfect and estimated channel knowledge. It can be seen that for perfect channel knowledge, the best possible result is reached after two iterations for this nonlinearity for both, 16-QAM and 64-QAM. However, if the channel is estimated, in both cases there remains a significant performance gap to the optimal case and there is even an error floor.

The result is similar for a more severe TWTA-type nonlinearity as shown in Figure 4.13. In the case of perfect channel knowledge, the optimal performance is eventually reached. For estimated channel knowledge however, there is a performance gap and an error floor in the high SNR region.

The reason for this behaviour is independent of the blind nonlinearity estimation since a similar result shows when the parameters of the nonlinearity are assumed as perfectly known. An erroneous channel estimation limits the quality of the correction signal that is subtracted in each iteration. This in turn leads to residual noise on

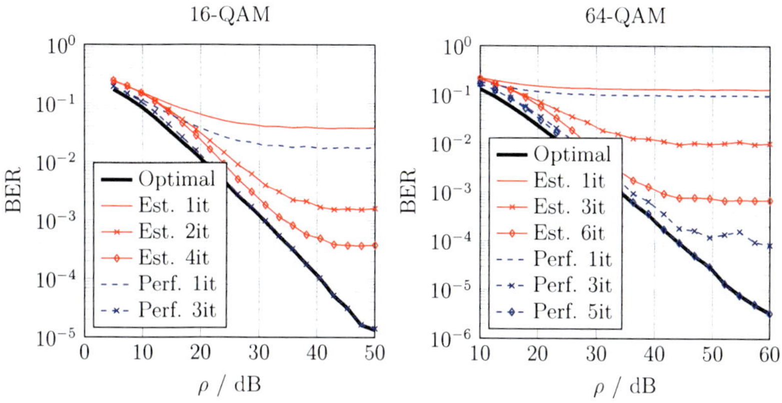

Figure 4.13.: BER comparison for TWTA nonlinearity with perfect and estimated channel knowledge, $\beta_A = 0.3, \alpha_\phi = \beta_\phi = 2$

the pilot carriers which limits channel estimation. After some iterations, the system converges into a non-optimal state. It can therefore be said that *in order to achieve high performance from this system, the channel and nonlinearity estimation need to be carefully designed with respect to their interdependencies.*

4.5. Conclusions

In this chapter, a feedback based iterative blind estimation algorithm for AM/AM and AM/PM nonlinearities has been presented. It was shown that by assuming perfect feedback in each iteration, the computational complexity for the parameter estimation is dramatically reduced. This gives the feedback based estimator a huge advantage over the feed forward based estimator especially when estimating AM/PM nonlinearities. Estimating those in a feed forward way is prohibitively expensive and suboptimal algorithms introduce large errors in the low SNR region. The feedback based estimator has none of these problems.

The requirements on the receiver are also very low. It has shown that in many cases, just one iteration is enough to get a very good estimation result. Since the received bits are available in almost all receiver architectures at some point, adding the feedback based estimator should be possible in the majority of cases. The ability to use feedback makes the algorithm a good fit to use in today's very popular turbo decoder and equalizer architectures.

Due to the feedback the algorithm is able to surpass the CRLB of the feed forward MLE and hence the theoretical best performance of the feed forward estimation method. These results make the feedback based estimator superior to the feed forward based methods and should be used if possible.

A more practical simulation employing an imperfectly estimated frequency selective channel revealed a strong interdependency between the performance of nonlinearity estimation and channel estimation. Both estimators could improve their performance iteratively based on the results of the other. This shows the dependency of the iterative estimator from the other components and the need for careful system design when employing this technique.

Lastly, there are strong indicators that the method can not only be used for OFDM signals but also for next generation Generalized Frequency Division Multiplexing (GFDM) or Filter Bank Multi Carrier (FBMC) schemes and might even be usable for single carrier modulated signals. However, some investigation is required to ensure that this is the case.

5. Real-Time implementation of blind Amplitude to Amplitude (AM/AM) estimation

5.1. Introduction

In the past chapters both feed forward and feedback based methods for the blind estimation of memoryless nonlinearities have been investigated. Two of these methods show to be computationally very inexpensive while at the same time offering a high degree of estimation performance. These are the blind feed forward AM/AM estimation and the iterative feedback based joint AM/AM and Amplitude to Phase (AM/PM) estimation. The iterative feedback based estimator has already been analyzed in a simulation scenario without perfect channel knowledge showing the high interdependency between the estimator and other system components.

After theoretical derivations and simulations, the next step is the implementation and testing of wireless communications algorithms on actual radio hardware in order to analyze the suitability of the algorithm for real world scenarios. To achieve that, the feed forward estimator for AM/AM nonlinearities, as described in chapter 3, is implemented on an Software Defined Radio (SDR) platform to investigate its performance with real nonlinearities which do not fit the estimation model perfectly [DF12]. The purpose of this implementation is to answer the following three questions.

Firstly, is the algorithm robust enough to cope with additional influences in a real communications system such as impairments in the analog frontend, imperfect channel knowledge et cetera? Secondly, is the complexity low enough so that a real-time implementation on standard hardware is possible? And finally, is the algorithm able to provide significant performance gains not only for perfectly memoryless nonlinearities but also for real amplifiers operating at the edge of saturation mode and which do not perfectly fit the assumed models?

5.2. System specification

Figure 5.1 shows the block diagram of the transmitter. It is an Orthogonal Frequency Division Multiplexing (OFDM) transmitter that supports Quaternary Phase Shift Keying (QPSK) up to 256-QAM modulated data symbols. For channel estimation, pilots can be assigned to some subcarriers. Pilot carriers are modulated with pseudo-randomly generated QPSK symbols. For robustness against frequency selectivity and for synchronization purposes a cyclic prefix is inserted after each OFDM transmit symbol $\boldsymbol{S}$ has been transformed to the time domain. However, the system is designed for laboratory tests of the nonlinearity mitigation and hence for nearly-flat channels with small delay spread. There are multiple ways of introducing nonlinearity into the system. As a reference, a virtual Solid State Power Amplifier (SSPA)-type nonlinearity can be enabled in the transmitter and arbitrarily parameterized. Furthermore, by setting the gain of the transmit and receiver amplifiers appropriately, nonlinear effects from real analog components can be introduced into the transmission.

The whole analog domain of the system is summarized by the Universal Software Radio Platform (USRP) blocks at the transmitter and receiver. Here, the digital signal is converted to the analog domain, pulse shaped, up-converted to the desired carrier frequency, amplified and finally transmitted via antenna or cable. On the receiver side, the signal is amplified, down-converted to baseband and transformed to the digital domain by sampling. The USRP is designed as a direct conversion architecture [NI].

Figure 5.2 shows the receiver's block diagram. Synchronisation is performed using an autocorrelation approach based on the cyclic prefix of the OFDM signal [SvdBB95]. After removal of the cyclic prefix, the signal is used for blind estimation of the nonlinearity as described in section 3.4. Furthermore, the pilots are extracted and used to reconstruct the amplitude and phase characteristic of the frequency selective channel. Afterwards, these effects are removed by means of zero forcing equalization. The signal is then demodulated and the estimated data is used for decision feedback mitigation of the nonlinearity as described in section 2.3.3.

In every mode of operation, an SSPA nonlinearity model is used for the blind estimation and mitigation. The model has not been tuned to match the actual hardware nonlinearities. This is to investigate the feasibility of the SSPA as a general model for nonlinearity estimation.

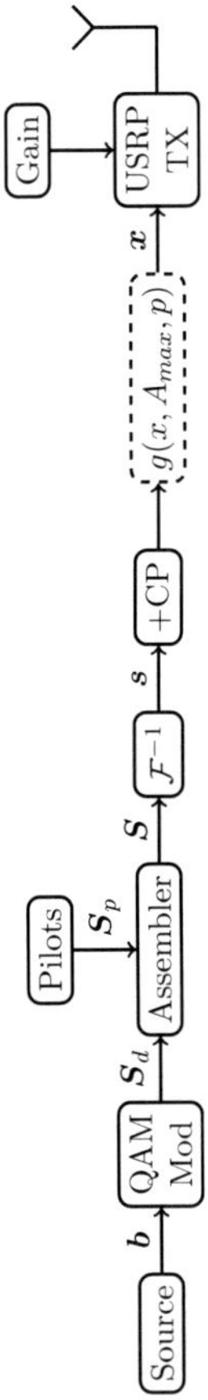

Figure 5.1.: Transmitter block diagram

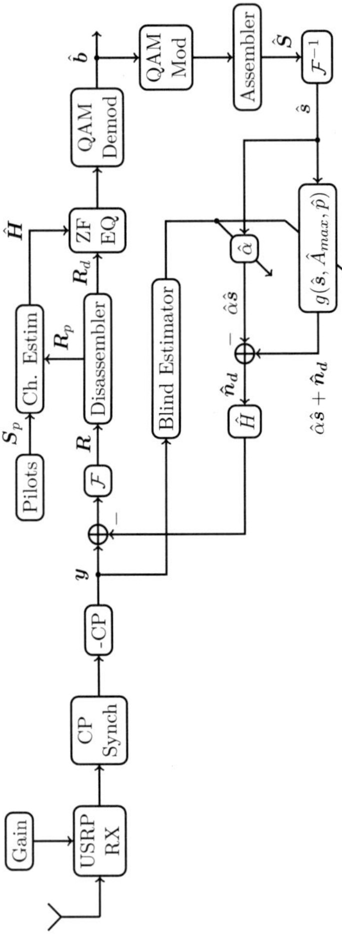

Figure 5.2.: Receiver block diagram

5.3. Low-complexity numerical convolution

All investigated feed forward estimation methods required a highly complex convolution to calculate the Probability Density Function (PDF) of the received signal after addition of Additive White Gaussian Noise (AWGN). When the samples are assumed to be mutually independent, the computational complexity in the general case is reduced to $O(L^2 \log L)$ when using 2D-Fast Fourier Transformation (FFT) based convolution and sampling with L samples per dimension (see section 3.4). However, for the given real-time implementation this is still a major issue and needs to be reduced further. Since in the given scenario, all densities are circular symmetric, a numerical convolution method with smaller complexity exists which is presented now.

5.3.1. 2D Fourier transform of circular symmetric functions

Let $\xi_1(x_0, x_1)$ and $\xi_2(x_0, x_1)$ be two circularly symmetric functions, i.e.,

$$\xi_1(x_0, x_1) = \xi_1(x) \tag{5.1}$$

$$\xi_2(x_0, x_1) = \xi_2(x)\,, \tag{5.2}$$

with $x = \sqrt{x_0^2 + x_1^2}$. Furthermore, let $\xi_3(x_0, x_1)$ be their convolution, i.e.

$$\xi_3(x_0, x_1) = \iint_{\infty}^{\infty} \xi_1(x_0', x_1') \cdot \xi_2(x_0 - x_0', x_1 - x_1')\mathrm{d}x_0'\mathrm{d}x_1'\,. \tag{5.3}$$

Using the Fourier transform, (5.3) can be calculated as

$$\xi_3(x_0, x_1) = \mathcal{F}^{-1}\left\{\mathcal{F}\left\{\xi_1(x_0, x_1)\right\} \cdot \mathcal{F}\left\{\xi_2(x_0, x_1)\right\}\right\} \tag{5.4}$$

which can be quickly calculated by means of the 2D-FFT as stated above. The 2D Fourier transform is defined as:

$$\Xi_1(u_0, u_1) = \iint_{-\infty}^{\infty} \xi_1(x_0, x_1)\mathrm{e}^{-j2\pi(x_0u_0 + x_1u_1)}\mathrm{d}x_0\mathrm{d}x_1\,. \tag{5.5}$$

It is well known that for a circularly symmatric function ξ_1, the Fourier transform Ξ_1 also exhibits circular symmetry [Bra99]. It can then be shown that (5.5) can be written as

$$\Xi_1(u) = 2\pi \int_0^{\infty} x \cdot \xi_1(x) \cdot J_0(2\pi u x)\mathrm{d}x\,, \tag{5.6}$$

where $u = \sqrt{u_0^2 + u_1^2}$ and J_0 denotes the zeroth order Bessel function of the first kind. This integral transform is known as Hankel transform [Bra99] and it can be said that a 2D Fourier transform degenerates into a 1D Hankel transform for a circularly symmetric function. With well designed lookup tables for the Bessel function, (5.6) could be implemented with a computational complexity of $O(L^2)$ which is already an improvement as compared to the general method but still not the least complex method.

5.3.2. The Fourier-Hankel-Abel cycle

The Fourier-Hankel-Abel cycle [Bra99] states that a 1-dimensional Hankel transform is equivalent to an Abel transform followed by a Fourier transform. Since the FFT is commonly accepted as the fastest general way for the Fourier transform, the focus here is on the Abel transform.

The Abel transform is defined as the projection of a circularly symmetric function onto an arbitrary axis [Bra99] and is given by the following equations:

$$\begin{aligned} \xi_1^A(R) &= 2\int_R^\infty \frac{\xi_1(x)R\mathrm{d}x}{\sqrt{x^2 - R^2}} \\ \xi_1(x) &= -\frac{1}{\pi}\int_x^\infty \frac{\xi_1^{A'}(R)\mathrm{d}R}{\sqrt{R^2 - x^2}}\,. \end{aligned} \tag{5.7}$$

In [HL85], Hansen describes an algorithm that approximates (5.7) by means of a state-variable model with negligible error. The implementation is extremely efficient and runs in linear time. By splitting up the desired Hankel transform into an Abel transform running in $O(L)$ and an FFT running in $O(L\log L)$, the complexity for the whole 2D-convolution can be reduced from $O(L^2\log L)$ to $O(L\log L)$ if both functions exhibit circular symmetry. Figure 5.3 shows the resulting complexities for several possible algorithms of the 2-dimensional convolution. In simulations $L = 1024$ has shown to be a good compromise of accuracy and speed so this value is used as the baseline for comparison. It can be seen that by exploiting the circular symmetry and transforming the 2D-FFT into a Hankel transform, a performance gain of factor 10 is possible. Splitting up the Hankel transform into an Abel and a Fourier transform then gains another factor 100 *yielding a factor 1000 performance gain as compared to the common 2D-FFT method.* These optimizations enabled the successful implementation of the method in real-time in the first place.

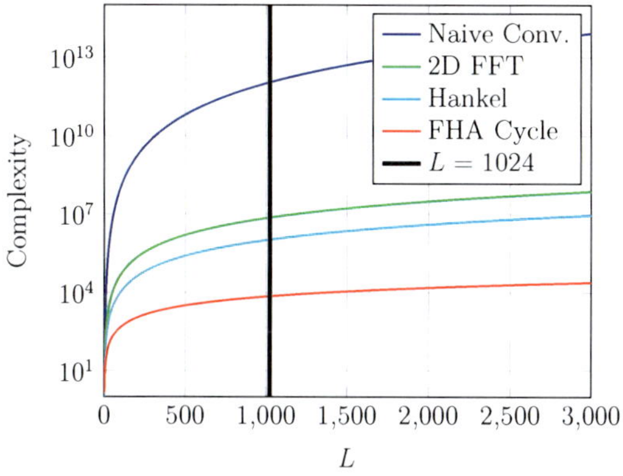

Figure 5.3.: Complexities of several 2-dimensional convolution algorithms

5.4. System Architecture

Physical Setup The demonstrator consists of a host Personal Computer (PC) running National Instruments LabVIEW [lab] and two USRP-2921 units [usr]. The physical setup and high level software architecture of the system is shown in Figure 5.4. The host PC is an off-the-shelf Intel Core i7 quadcore system with 3.4 GHz clock rate and 16 GByte of RAM. The complex baseband samples are streamed to and from the USRPs via dedicated gigabit Ethernet ports on the host machine.

High level software architecture Inside the LabVIEW environment, transmitter and receiver are running completely independently of one another so that the system can be easily split up and separated physically. The Graphical User Interface (GUI) is another separate process that shows the transmission status and is used to control the whole system. The transmission is independent of the GUI so it can be easily replaced by a less involved interface or shut down completely to free more resources.

5.4.1. Transmission parameters

Generally, the system is highly configureable and many of the following parameters can be changed. However, in the authors opinion the transmission parameters shown in table 5.1 give good results and run very stable. A high oversampling ratio is chosen because the lowpass filter after the Digital-to-Analog Converter (DAC) in

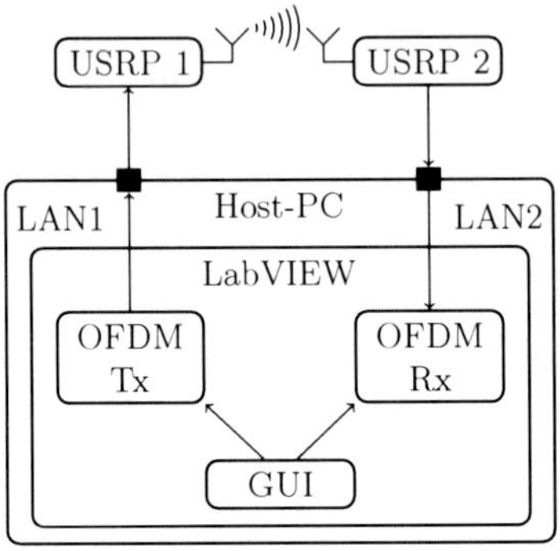

Figure 5.4.: Demonstrator physical setup and high level software architecture

Table 5.1.: OFDM Transmission Parameters

Parameter	Value
I/Q-Rate	8 MS/s
Oversampling	4-fold
Subcarriers	1024
Active carriers	764
Pilot carriers	191
Bandwidth / Carrier	1.95 kHz
Used Bandwidth	1.492 MHz
Carrier frequency	2.45 GHz
Modulation Schemes	QPSK 16-QAM 64-QAM

the USRPs induces significant amounts of Inter Carrier Interference (ICI) otherwise. By using 4-fold oversampling this problem is largely alleviated and without any nonlinear distortions, roughly 25 dB of Signal to Noise Ratio (SNR) at the input of the detector is achieved for a cable transmission. Given a powerful enough host machine, the demonstrator is able to adjust the subcarrier and hence the total system bandwidth with the change of one single parameter. On each side of the OFDM symbol there are 128 guardband carriers. Furthermore, four carriers around Direct Current (DC) are nulled. To comply with regulations, the licence free 2.4 GHz band is chosen. The specific carrier frequency of 2.45 GHz is selected because it had the least interferers in the author's environment and can be freely changed without notable influence on the transmission.

5.4.2. Transmitter Architecture

When measuring the performance, the transmitter initially generates 256 data packets, each containing bits for one OFDM symbol, using a pseudo random bit generator. During operation, one packet is selected at random, modulated and transmitted. By knowing the seed of the random bit generator, the receiver is able to generate the same set of packets and hence, no auxillary channel is required to transfer the data to the receiver for bit error counting. Due to the architecture, by replacing the random packet selector with a real information source, the demonstrator can be easily transformed into a true streaming system at the loss of Bit Error Rate (BER) statistics since the receiver will not have correct reference data to compare against.

5.4.3. Receiver Architecture

The receiver architecture is shown in Figure 5.5. To maximize the usage of the available cores it is split up into three processes that run independently and in parallel. This is ensured by using non-synchronous communications between them that does not require any waiting as long as there is data available to process. Large data packets, such as OFDM symbols that should not be lost, are transmitted using queues and notifiers are used for events such as new parameter estimates or a loss of synchronization where a new event makes the older one redundant. This architecture ensures loose coupling and hence supports parallel operation.

The stream of complex baseband samples from the USRP is processed by a

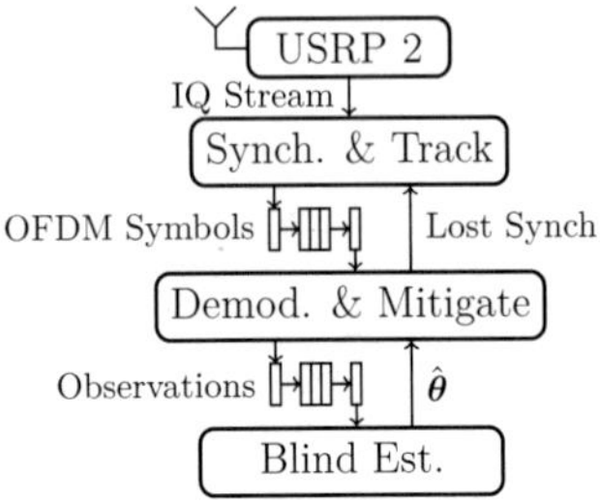

Figure 5.5.: Receiver Architecture

synchronization and tracking unit. Initially the unit synchronizes to the OFDM symbol clock and passes the time-domain samples belonging to one OFDM symbol to the next stage. A cyclic prefix based synchronization method is used. After synchronization this stage enters a tracking mode in which slowly accumulating sampling offsets are mitigated.

The demodulation and mitigation implements channel equalization and the decision feedback detector. This unit also calculates statistics such as SNR and BER. For nonlinearity mitigation it requires a valid estimate of the parameters $\boldsymbol{\theta}$ from the estimation unit. This stage can be configured to drop symbols when the input queue is filling up, thus allowing tests with very large sampling rates that would otherwise lead to buffer overflows.

Finally the last unit performs the blind AM/AM estimation assuming a SSPA-type nonlinearity. It does not estimate per-symbol but aggregates a configurable amount of symbols. This allows a more precise estimation and reduces the computational load. A value of 20 symbols per estimation has shown to be sufficient in order to avoid buffer overflows in the input queue.

5.5. Results

A photo of the demonstrator setup without the host PC is shown in Figure 5.6 and a screenshot of the GUI in Figure 5.7. In the following, two cases are investigated. For reference, the gains are chosen such so that the hardware amplifiers do not introduce nonlinear distortion. Then, the signal is distorted by the transmitter using a virtual SSPA nonlinearity. Hence, the estimation and mitigation model matches the actual nonlinearity perfectly which gives the best possible results. In a second trial, the

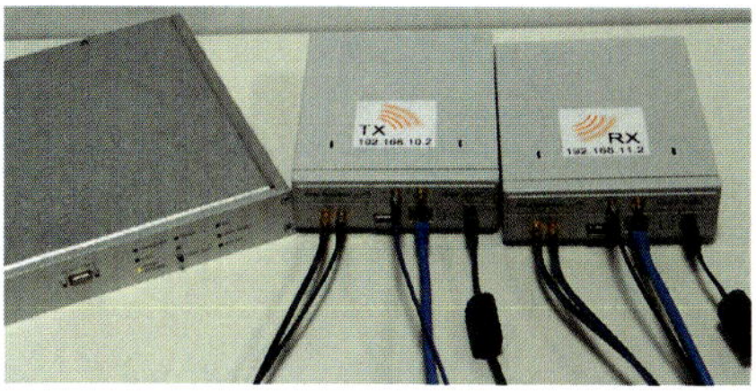

Figure 5.6.: Demonstrator setup

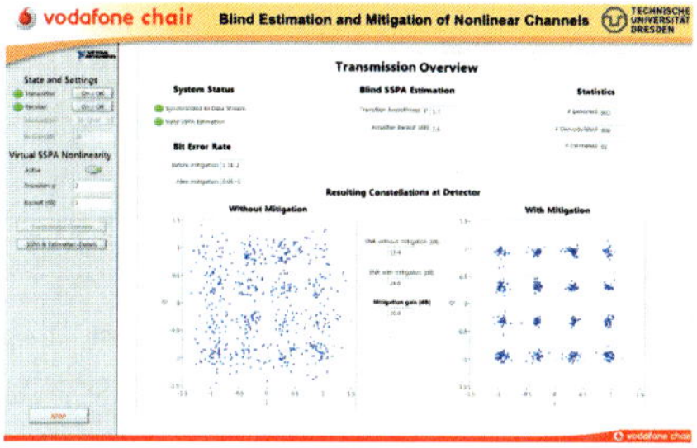

Figure 5.7.: GUI Screenshot

gain of the receiver amplifier is steadily increased to investigate the performance of the algorithm with an actual hardware nonlinearity. In all cases, only one iteration of the decision feedback receiver is utilized.

5.5.1. Simulated SSPA Nonlinearity

Table 5.2 shows the results for the simulated SSPA nonlinearity. Very high gains in terms of SNR and BER are achieved as predicted by the earlier simulation results. This mainly shows the robustness of the estimation and mitigation against the additional impairments of an actual radio transmission.

5.5.2. USRP Receiver Nonlinearity

Table 5.3 shows the results when the amplifier of the receiver USRP is driven into saturation. SNR gains of 3-4 dB are possible over a wide range of gain settings. While QPSK is relatively robust against the induced nonlinear distortions and suffers only from very few errors in the case of more severe distortion, for 16-QAM gains of

Table 5.2.: Results for Simulated SSPA Nonlinearity

Scenario		SNR[dB] / BER		
Modulation	p / γ[dB]	No mitigation	Mitigation	SNR Gain
QPSK	Linear	25.0 / 0	N/A	N/A
16-QAM	Linear	25.0 / 0	N/A	N/A
QPSK	2 / 3	15.7 / 0	23.4 / 0	7.7
QPSK	1 / 2	13.9 / 0	24.7 / 0	10.8
QPSK	1 / 0	12.2 / 4.0e-6	24.6 / 0	12.4
16QAM	2 / 3	15.6 / 2.2e-3	22.8 / 0	7.2
16QAM	1 / 2	14.1 / 8.2e-3	21.7 / 1.5e-4	7.6
16QAM	1 / 0	12.4 / 2.0e-2	19.1 / 5.0e-3	6.7

one to two orders of magnitude in terms of BER can be observed. These results show a significant improvement of performance in the presence of a hardware nonlinearity.

One explaination for the difference between simulated and hardware amplifier is a model mismatch. Figure 5.8 shows the densities of the incoming samples and the expected density for the estimated nonlinearity. Clearly, the hardware nonlinearity

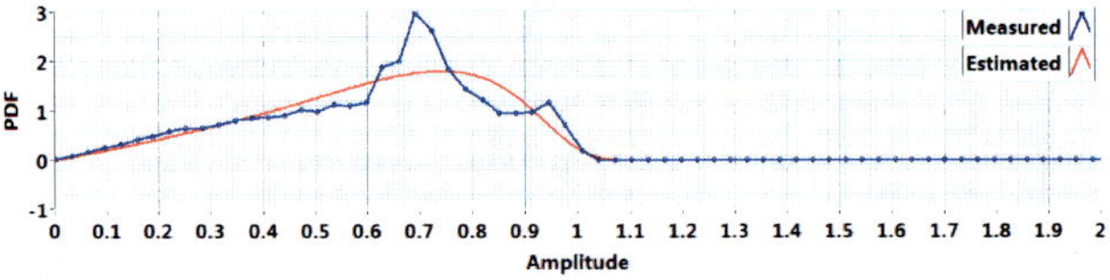

Figure 5.8.: Misestimation of Amplifier Characteristic

exhibits features that cannot be adequately represented by the SSPA model. On the other hand, despite this model mismatches, significant performance gains are achieved. In the end, there is a tradeoff between a simple model that can be more easily estimated but exhibits some mismatches, a more complicated model that requires higher estimation efforts and a specialized model that may not be a good fit for every amplifier characteristic.

Table 5.3.: Results for overdriven USRP receiver

Scenario		SNR[dB] / BER		
Modulation	Gain[dB]	No mitigation	Mitigation	SNR Gain
QPSK	70	26.1 / 0e0	N/A	N/A
QPSK	75	21.0 / 0e0	23.0 / 0e0	2.0
QPSK	77	16.1 / 0e0	20.4 / 0e0	4.3
QPSK	80	11.0 / 8e-5	17.7 / 0e0	6.7
16QAM	70	26.0 / 0e0	N/A	N/A
16QAM	75	19.3 / 1e-4	22.2 / 1e-6	2.9
16QAM	76	16.8 / 1e-3	20.7 / 1e-5	3.9
16QAM	77	15.0 / 5.5e-3	18.7 / 2.5e-4	3.7
16QAM	78	13.5 / 1.5e-2	17.8 / 2.4e-3	4.3
16QAM	80	10.0 / 5e-2	13.2 / 3e-2	3.2

5.6. Conclusions

In this chapter, two implementations of blind estimation algorithms in practical scenarios are investigated.

The first demonstrator is implemented on a SDR platform. It shows that using a generic model with just two parameters, the nonlinearity of an overdriven hardware amplifier can be estimated such that performance gains of 3-4 dB in terms on SNR and one to two orders of magnitude in terms of BER are possible. The system also proves to be tolerant to mismatches between the assumed model and actual nonlinearity. This indicates that simple generic nonlinearity models can be used to estimate a wide range of hardware nonlinearities.

Secondly the iterative blind estimator is implemented under the assumption of imperfect channel knowledge. While the blind estimation shows to be good enough to not be a bottleneck, there is a strong interdependence between reachable performance and quality of the channel estimates. It shows that the combination of decision feedback mitigation and imperfect channel estimation can lead the system to converge to a non-optimal result. Therefore, a very good channel estimation algorithm needs to be employed when implementing blind nonlinearity mitigation.

6. Conclusions

The following goals for this work have been defined in the introduction.

1. Show the feasibility of blind nonlinearity estimation by mathematical derivation of estimators and characterization of their performance.
2. Analyze the complexity of the estimators and evaluate their applicability to use in real-time environments. If required, derive low-complexity algorithms and characterize the tradeoffs.
3. Show that the resulting algorithms are suitable for use in real world scenarios by testing them with actual hardware amplifiers operated in saturation mode.

The results are summarized in the following four theses.

Thesis 1: Blind nonlinearity estimation is possible

The general maximum likelihood estimator for the joint estimation of an Amplitude to Amplitude (AM/AM) and Amplitude to Phase (AM/PM) type nonlinearity with a number of unknown parameters has been derived in a feedforward manner. While it proved to be prohibitively complex to be implemented on real hardware, it indicates the possibility of blind nonlinearity estimation in general. For the important special case of independent samples and AM/AM distortion only, the Cramér-Rao Lower Bound (CRLB) was also derived and it was shown that the ML-estimator reaches it. This case has subsequently been extended to cover non-independent samples and frequency selective channel scenarios with controllable performance loss. Furthermore, another special case allowing the estimation of AM/PM nonlinearities has also been derived and shown to provide satisfying performance for weak AM/PM nonlinearities.

Furthermore, an iterative feedback estimator has been derived. In all investigated cases it reached or even bested the performance of the feedforward based algorithm.

It can be implemented to jointly estimate amplitude and phase distortions. It is not limited to Orthogonal Frequency Division Multiplexing (OFDM). Instead, it possibly covers a wide range of transmission models and hence is very generic and powerful.

Thesis 2: Blind nonlinearity estimation is computationally feasible

The special case of feedforward AM/AM estimation was reduced to $O(L \log L)$ complexity and hence, is no more complex than a standard Fast Fourier Transformation (FFT). This makes the algorithm readily implementable on today's common hardware. This has been proven by the development of a real-time demonstrator implementing this algorithm which runs on a basic off-the-shelf Personal Computer (PC) and two low-cost software defined radios.

For the feedforward AM/PM estimation a low-complexity algorithm has been shown as well. However, it suffers from severe bias in the low Signal to Noise Ratio (SNR) region. Nevertheless, for weak nonlinearities it provides nearly optimal performance, as well.

Iterative estimation has shown to be extremely low on computational complexity. Generally, implementation in $O(L \log L)$ is possible, again stemming from the required FFTs. Since it allows to jointly estimate amplitude and phase distortions, it is usually superior to the feedforward based methods in terms of overall complexity.

Thesis 3: The blind estimators are compatible with many receiver architectures

For nonlinearity estimation, the requirements imposed on the receiver are small. For the feedforward methods, merely the raw baseband samples belonging to the OFDM symbols and a channel estimate is required. This is available in a wide range of receivers.

For the iterative estimation, only the feedback signal is required in addition. Since the received bits are available at some point in almost any receiver, remodulation can usually be performed if the hardware has enough computational reserves.

Thesis 4: Blind estimation is feasible in real radio scenarios

A blind estimation and mitigation scheme has been implemented on a software radio platform in real time. It was possible to obtain a significant 3-4 dB gain in SNR and one to two orders of magnitude in terms of Bit Error Rate (BER) for an overdriven hardware amplifier without tailoring the algorithm to this specific nonlinearity. This shows the generality of the approach and the feasibility to operate in actual communications systems.

Furthermore the behaviour in a scenario with imperfect channel knowledge was investigated. It has been shown that the estimation does not suffer significantly but the combination of imperfect nonlinearity mitigation and channel estimation can lead to non-optimal performance. Yet, significant gains compared to linear reception have been achieved and with careful system design those problems can be largely alleviated.

6.1. Future work

The goal of this thesis, proving that blind estimation of non-linearities is possible and leads to significant performance gains, has been fulfilled. However, several open questions remain to be answered.

1. The blind iterative estimator shows potential to be usable in a much wider range of scenarios as those covered in this thesis. For example, it may be applicable to 5G modulation schemes such as Generalized Frequency Division Multiplexing (GFDM) or Filter Bank Multi Carrier (FBMC) and possibly even single carrier systems. It should be investigated if this is actually the case and how well the estimator performs.

2. It was shown that imperfect channel estimation severely degrades performance of nonlinearity estimation. This was presented using a very basic channel estimation and equalization scheme. The influence of different estimators and equalizers on the performance of blind estimation needs to be investigated.

3. Since the decision feedback detector and iterative estimator harnesses feedback information, they appear to be a good fit with other iterative methods such

as turbo coding and turbo equalization. The resulting performance in a fully iterative receiver should be looked into.

4. So far, only the feed forward based AM/AM estimator has been tested with actual hardware nonlinearities yielding significant performance gains. It was shown in the thesis that the feedback based estimation approach outperforms the feedforward based approach. Hence, it should be implemented and tested with real amplifiers to verify this results in practise.

5. All of the estimation methods have been derived for a single-antenna system only. Systems with multiple antennas are becoming more and more common nowadays. Since each additional antenna usually means at least another power amplifier, multi antenna systems would benefit even more from reduced requirements on the amplifiers. Hence, the methods should be adapted and tested with multiple antennas.

6. The methods have mainly been investigated for the case of a clipping amplifiers. However, mathematically they are not limited to those cases and can, for example, also estimate and mitigate other nonlinear components such as analog-to-digital converters. It should be investigates how useful the approaches presented in this thesis are for nonlinearities not originating in amplifiers.

7. It has been briefly mentioned, that non-linear distortions also cause out-of-band radiation which might lead to the violation of spectral masks. There exist a number of transmitter based algorithms that reduce these emissions but often leave residual effects. A real system study containing transmitter and receiver based methods is a neccessary step to prove the feasibility of these methods in the real world.

A. Random Variable Transformations

Random variable transformations were used in chapter 3 of this thesis to calculate the Probability Density Function (PDF)s of random variables after being subject to nonlinear distortions and other operations. Essentially, given a random variable $\boldsymbol{x}$ with known PDF $p(\boldsymbol{x})$ and a deterministic function $\boldsymbol{y} = g(\boldsymbol{x})$, random variable transformations are used to determine the PDF $p(\boldsymbol{y})$.

This chapter gives a short overview of the different types of random variable transformations. It is written to provide a general understanding of the topic and may trade mathematical rigor for vividness. A very detailled and rigorous introduction to the topic is given in [Pap91].

A.1. Invertible transformation of one variable

One of the simplest cases is the transformation of a single random variable x through an invertible function $y = g(x)$, i.e. there exists $g^{-1}(y)$ so that for all x:

$$x = g^{-1}(y)\,. \tag{A.1}$$

Furthermore, the first derivative of g must exist, hence:

$$\frac{\mathrm{d}y}{\mathrm{d}x} = \frac{\mathrm{d}g(x)}{\mathrm{d}x} = g'(x)\,. \tag{A.2}$$

The principle is shown in Figure A.1. The function $g(x)$ maps every $x_0 \in \boldsymbol{X}$ to an element $y_0 \in \boldsymbol{Y}$. Since the function is invertible, there exists a reverse mapping from y_0 to x_0 through the inverse function $g^{-1}(y)$.

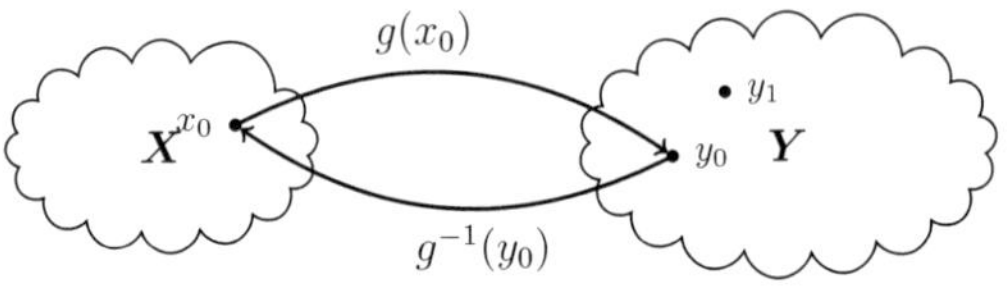

Figure A.1.: Invertible mapping between $\boldsymbol{X}$ and $\boldsymbol{Y}$

Determining the resulting PDF It is known that the probability of y to take a value in a very small environment $\mathrm{d}y$ around y_0 is given as follows:

$$Pr\left(y_0 < y \leq y_0 + \mathrm{d}y\right) = p_y(y_0) \cdot \mathrm{d}y\,. \tag{A.3}$$

Similarly, we can write:

$$Pr\left(g^{-1}(y_0) < x \leq g^{-1}(y_0) + \frac{\mathrm{d}y}{g'(y_0)}\right) = p_x(x_0) \cdot \mathrm{d}x\,. \tag{A.4}$$

For $x_0 = g^{-1}(y_0)$, the respective probabilites are equal, hence:

$$p_y(y_0) \cdot \mathrm{d}y = p_x(g^{-1}(y_0)) \cdot \mathrm{d}x\,. \tag{A.5}$$

Since the derivative $\mathrm{d}y/\mathrm{d}x = g'(x)$ is known, we finally arrive at the well known formula:

$$p_y(y_0) = \frac{p_x(g^{-1}(y_0))}{|g'(g^{-1}(y_0))|}\,. \tag{A.6}$$

The absolute value in the denominator is required because in Equation (A.4) it was implicitely assumed that g' is positive. More details on that can be found in [Pap91].

From a practical point of view it is often convenient to allow elements $y_1 \in \boldsymbol{Y}$ for which there is no x_1 such that $y_1 = g(x_1)$. For example, an amplifier that limits the output amplitude to a certain maximum value y_{max}. When using the resulting PDFs, e.g. in simulations or parameter estimation, they should be well defined for $y > y_{max}$. Since those values can never appear at the output of the amplifier, their associated probability is zero and hence, the PDF at these values is zero as well.

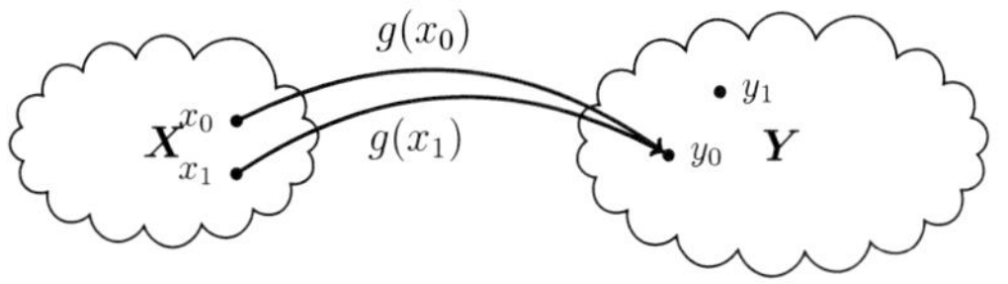

Figure A.2.: Non-invertible mapping between $\boldsymbol{X}$ and $\boldsymbol{Y}$

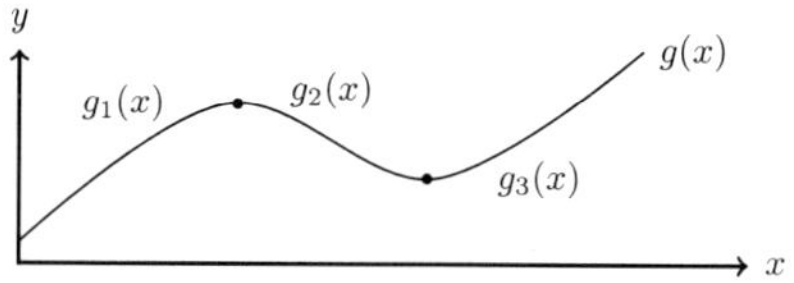

Figure A.3.: Decomposition of $g(x)$ into invertible functions

A.2. Non-invertible transformation of one variable

While the former already covers a lot of cases found in practise, non-invertible functions are very common as well, e.g. the AM/AM characteristic of a Travelling Wave Tube Amplifier (TWTA) amplifier. The principle is shown in Figure A.2. Now, multiple input values can be mapped to a single output value. Hence, there exists no invertible function $g^{-1}(y)$. However, intuitively the problem should be solveable since calculating the probability of observing y_0 at the output is just the sum of the probability of either x_0 or x_1 occuring at the input.

For this example, the scope is limited to piecewise monotonic functions. Hence, $g(x)$ can be decomposed into N invertible functions $g_1(x), ..., g_N(x)$ with non-overlapping domains. This is shown in Figure A.3. For each part, the inverse function $g_k^{-1}(y)$ exists.

Determining the resulting PDF Since the probability of y taking a value in a small environment around y_0 is the sum of all possible probabilies of x falling in a small area around $g_1^{-1}(y_0), ..., g_N^{-1}(y_0)$ it can be concluded that:

$$p_y(y) = \sum_{k=1}^{N} \frac{p_x(g_k^{-1}(y))}{g'(g_k^{-1}(y))} . \tag{A.7}$$

It should be noted that there might exist values y_0 for which some $g_k^{-1}(y_0)$ are not defined. This is the case when the monotonic piece $g_k(x)$ contains no mapping such that $g_k(x) = y_0$. In those cases, the summands become zero. Special care must be taken of this when implementing the calculation of such probability density functions.

A.3. Transformation of Random Vectors

A very common case is when $\boldsymbol{x}$ is an m-dimensional random vector instead of a scalar random variable. The transformation is given by the function $\boldsymbol{y} = g(\boldsymbol{x})$, $g : \mathbb{R}^m \to \mathbb{R}^m$. It is assumed that the inverse $\boldsymbol{x} = g^{-1}(\boldsymbol{y})$ exists and similar ideas to that given in the last section can be applied if not.

Determining the resulting PDF The line of thought is very similar to the derivation given in section A.1 and hence, the resulting equation is very similar:

$$p_y(\boldsymbol{y_0}) = \frac{p_x(g^{-1}(\boldsymbol{y_0}))}{|J(g^{-1}(\boldsymbol{y_0}))|}. \tag{A.8}$$

$J(\boldsymbol{x})$ is the Jacobian determinant which is defined as

$$J(\boldsymbol{x}) = \begin{vmatrix} \frac{\partial y_0}{\partial x_0} & \cdots & \frac{\partial y_0}{\partial x_{m-1}} \\ & \vdots & \\ \frac{\partial y_{m-1}}{\partial x_0} & \cdots & \frac{\partial y_{m-1}}{\partial x_{m-1}} \end{vmatrix}. \tag{A.9}$$

Again, for more details the reader is pointed to [Pap91].

Bibliography

[AD12] Teresa Araujo and Rui Dinis. On the Accuracy of the Gaussian Approximation for the Evaluation of Nonlinear Effects in OFDM Signals. *IEEE Transactions on Communications*, 60(2):346–351, February 2012.

[Arm02] J Armstrong. Peak-to-average power reduction for OFDM by repeated clipping and frequency domain filtering. *Electronics letters*, 38(5):2001–2002, 2002.

[BFH96] RW Bauml, RFH Fischer, and JB Huber. Reducing the peak-to-average power ratio of multicarrier modulation by selected mapping. *Electronics Letters*, pages 2–6, 1996.

[BMWH01] H. Breiling, S.H. Muller-Weinfurtner, and J.B. Huber. SLM peak-power reduction without explicit side information. *IEEE Communications Letters*, 5(6):239–241, June 2001.

[Bra99] Ronald N Bracewell. *The Fourier Transform and its Applications*, volume McGraw-Hil of *McGraw-Hill series in electrical and computer engineering. ~Circuits and systems)*. McGraw-Hill, 1999.

[BS11] I Bashir and RB Staszewski. Autonomous predistortion calibration of an RF power amplifier. *Circuits and Systems (ISCAS), 2011 ...*, pages 205–208, 2011.

[BSE04] Ahmad R.S. Bahai, Burton R. Saltzberg, and Mustafa Ergen. *Multi-Carrier Digital Communications: Theory and Applications of OFDM (Information Technology: Transmission, Processing and Storage)*. Springer, 2004.

[Bus52] J J Bussgang. Crosscorrelation functions of amplitude-distorted Gaussian signals. Technical report, Research Lab. Electron, M.I.T., Cambridge, MA, Tech. Rep.216, 1952.

[BZF08] Steffen Bittner, Peter Zillmann, and Gerhard Fettweis. Iterative Correction of Clipped and Filtered Spatially Multiplexed OFDM Signals. *VTC Spring 2008 - IEEE Vehicular Technology Conference*, pages 953–957, May 2008.

[CK07] E. Ward Cheney and David R. Kincaid. *Numiercal Mathematics and Computing*. Brooks/Cole Publishing Co., 2007.

[CMP99] Elena Costa, Michele Midrio, and Silvano Pupolin. Impact of amplifier nonlinearities on OFDM transmission system performance. *Communications Letters, IEEE*, 3(2):37–39, 1999.

[CP02] Elena Costa and Silvano Pupolin. M-QAM-OFDM system performance in the presence of a nonlinear amplifier and phase noise. *Communications, IEEE Transactions on*, 50(3):462–472, 2002.

[CTL12] Tzi-Dar Chiueh, Pei-Yun Tsai, and I-Wei Lai. *Baseband Receiver Design for Wireless MIMO-OFDM Communications*. Wiley-IEEE Press, 2012.

[D'A01] AN D'Andrea. Nonlinear predistortion of OFDM signals over frequency-selective fading channels. *..., IEEE Transactions on*, 49(5):837–843, 2001.

[DF11] Jan Dohl and Gerhard Fettweis. Blind Estimation and Mitigation of Nonlinear Channels. In *Personal, Indoor and Mobile Radio Communications, 2011. PIMRC 2011. IEEE 22nd International Symposium on*, 2011.

[DF12] Jan Dohl and Gerhard Fettweis. Evaluation of Estimation and Mitigation Algorithms for Nonlinearly Distorted OFDM Signals on a SDR Platform. In *International Workshop on Signal Processing Advances in Wireless Communications (SPAWC'12)*, 2012.

[DF13] Jan Dohl and Gerhard Fettweis. Blind Estimation of Memoryless AM / PM Nonlinearities in OFDM Systems. In *Vehicular Technology Conference (VTC 2013-Fall)*, 2013.

[DF14] Jan Dohl and Gerhard Fettweis. Iterative blind estimation of nonlinear channels. *2014 IEEE International Conference on Acoustics, Speech and Signal Processing (ICASSP)*, pages 3923–3927, May 2014.

[DOB+09] Erik Dahlman, Claude Oestges, Alan C. Bovik, Bruce A. Fette, Keith Jack, Farid Dowla, Stefan Parkvall, Johan Skold, Casimer DeCusatis, Ed da Silva, Ron Olexa, Bruno Clerckx, Luis M. Correia, Philip A Chou, Mihaela van der Schaar, W. K. Ling, Ronald Kitchen, Daniel M. Dobkin, Dan Bensky, David Morgan, Juanita Ellis, Charles Pursell, Joy Rahman, Leonidas Guibas, and Feng Zhao. *Communications Engineering Desk Reference*. Academic Press, 2009.

[DS07] C Dehos and T C W Schenk. Digital Compensation of Amplifier Nonlinearities in the Receiver of a Wireless System. In *Communications and Vehicular Technology in the Benelux, 2007 14th IEEE Symposium on*, pages 1–6, 2007.

[DTV00] D Dardari, V Tralli, and A Vaccari. A theoretical characterization of nonlinear distortion effects in OFDM systems. *Communications, IEEE Transactions on*, 48(10):1755–1764, 2000.

[FNL12] G. Fettweis, W. Nagel, and W. Lehner. Pathways to servers of the future. In *2012 Design, Automation & Test in Europe Conference & Exhibition (DATE)*, pages 1161–1166. IEEE, March 2012.

[FuHLF13] Gerhard P. Fettweis, Najeeb ul Hassan, Lukas Landau, and Erik Fischer. Wireless Interconnect for Board and Chip Level. In *Design, Automation & Test in Europe Conference & Exhibition (DATE), 2013*, pages 958–963, New Jersey, 2013. IEEE Conference Publications.

[Gar07] Vijay Garg. *Wireless Communications & Networking (The Morgan Kaufmann Series in Networking)*. Morgan Kaufmann, 2007.

[HHH10] Christian Hofbauer, Mario Huemer, and Johannes B. Huber. Coded OFDM by unique word prefix. In *2010 IEEE International Conference on Communication Systems*, pages 426–430. IEEE, November 2010.

[HL85] Eric W. Hansen and Phaih-Lan Law. Recursive methods for computing the Abel transform and its inverse. *Journal of the Optical Society of America A*, 2(4):510, April 1985.

[JK03] Jiho Jang and Kwang-Bok Lee. Transmit power adaptation for multiuser OFDM systems. *IEEE Journal on Selected Areas in Communications*, 21:171–178, 2003.

[JT00] ADS Jayalath and C Tellambura. Reducing the peak-to-average power ratio of orthogonal frequency division multiplexing signal through bit or symbol interleaving. *Electronics Letters*, 36(13):1161–1163, 2000.

[Kay93a] Steven M Kay. Fundamentals of Statistical Signal Processing: Estimation Theory. *Technometrics*, 37:465, 1993.

[Kay93b] Steven M. Kay. *Fundamentals of Statistical Signal Processing, Volume I: Estimation Theory*. Prentice Hall, 1993.

[KDJ05] Don Kimball, Paul Draxler, and Jinho Jeong. 50% PAE WCDMA basestation amplifier implemented with GaN HFETs. In *CSIC'05*, pages 89–92, 2005.

[lab] http://www.ni.com/labview.

[LRWW98] JC Lagarias, JA Reeds, MH Wright, and PE Wright. Convergence properties of the Nelder–Mead simplex method in low dimensions. *SIAM Journal on Optimization*, 9(1):112–147, 1998.

[MBFH97] Stefan Müller, Robert Bäuml, Robert Fischer, and Johannes Huber. OFDM with reduced peak-to-average power ratio by multiple signal representation. *Annals of Telecommunications*, 52:58–67, 1997.

[MH97] SH Muller and JB Huber. OFDM with reduced peak-to-average power ratio by optimum combination of partial transmit sequences. *Electronics letters*, 33(5):368–369, 1997.

[Min85] John Minkoff. The role of AM-to-PM conversion in memoryless nonlinear systems. *Communications, IEEE Transactions on*, C(2):139–144, 1985.

[Nee99] Richard Van Nee. *OFDM for Wireless Multimedia Communications (Artech House Universal Personal Communications)*. Artech House Publishers, 1999.

[NI] Ni whitepaper: What is ni usrp hardware? http://www.ni.com/white-paper/12985/en/pdf.

[Nyq28] H. Nyquist. Certain Topics in Telegraph Transmission Theory. *Transactions of the American Institute of Electrical Engineers*, 47(2):617–644, April 1928.

[Pap91] Athanasios Papoulis. *Probability, Random Variables and Stochastic Processes*. McGraw-Hill Education (ISE Editions), 3rd edition, 1991.

[PM05] JC Pedro and SA Maas. A comparative overview of microwave and wireless power-amplifier behavioral modeling approaches. *Microwave Theory and Techniques, IEEE ...*, 53(4):1150–1163, 2005.

[PS07] John Proakis and Masoud Salehi. *Digital Communications, 5th Edition*. McGraw-Hill Science/Engineering/Math, 2007.

[Rap91] C Rapp. Effects of HPA-nonlinearities on a 4-DPSK/OFDM-signal for a digital sound broadcasting system. In *Proc. second European conf. on Satellite Commun., Liege, Belgium*, 1991.

[RC00] W. Rhee and J.M. Cioffi. Increase in capacity of multiuser OFDM system using dynamic subchannel allocation. *VTC2000-Spring. 2000 IEEE 51st Vehicular Technology Conference Proceedings (Cat. No.00CH37026)*, 2, 2000.

[Ryu11] HG Ryu. Combination of PAPR reduction and linearization for the OFDM communication system. *Wireless Communications and Mobile Computing*, (February 2010):46–52, 2011.

[RZ02] R. Raich and GT Zhou. On the modeling of memory nonlinear effects of power amplifiers for communication applications. In *Digital Signal Processing Workshop, 2002 and the 2nd Signal Processing Education Workshop. Proceedings of 2002 IEEE 10th*, pages 7–10, 2002.

[RZF06] Wolfgang Rave, Peter Zillmann, and Gerhard Fettweis. Iterative correction and decoding of OFDM signals affected by clipping. In *5th International Workshop on Multi-Carrier Systems and Solutions*, 2006.

[Sal81] A Saleh. Frequency-independent and frequency-dependent nonlinear models of TWT amplifiers. *Communications, IEEE Transactions on*, C(11), 1981.

[SDMF07] Tim C. Schenk, Cedric Dehos, Dominique Morche, and Erik Fledderus. Receiver-Based Compensation of Transmitter-Incurred Nonlinear Distortion in Multiple-Antenna OFDM Systems. *2007 IEEE 66th Vehicular Technology Conference*, pages 1346–1350, September 2007.

[SKH12] KY Son, Bonhoon Koo, and Songcheol Hong. A CMOS power amplifier with a built-in RF predistorter for handset applications. *Microwave Theory and Techniques, ...*, 60(8):2571–2580, 2012.

[SSB09] AMJ Schoonen, T.C.W. Schenk, and P.G.M. Baltus. Impact and digital estimation of nonlinear impairments in analog baseband processing stages. In *Wireless Technology, 2008. EuWiT 2008. European Conference on*, pages 73–76. IEEE, 2009.

[SvdBB95] Magnus Sandell, J.J. van de Beek, and P.O. Börjesson. Timing and frequency synchronization in OFDM systems using the cyclic prefix. In *Proc. Int. Symp. Synchronization*, number December, pages 14–15. Citeseer, 1995.

[THC03] Jose Tellado, L.M.C. Hoo, and J.M. Cioffi. Maximum-likelihood detection of nonlinearly distorted multicarrier symbols by iterative decoding. *Communications, IEEE Transactions on*, 51(2):218–228, 2003.

[usr] http://sine.ni.com/nips/cds/view/p/lang/en/nid/209949.

[WJ65] John M. Wozencraft and Irwin Mark Jacobs. *Principles of communication engineering.* 1965.

[WL11] Y Wang and Z Luo. Optimized iterative clipping and filtering for PAPR reduction of OFDM signals. *Communications, IEEE Transactions on*, 59(1):33–37, 2011.

[WPP08] KD Wong, MO Pun, and HV Poor. The continuous-time peak-to-average power ratio of OFDM signals using complex modulation schemes. *..., IEEE Transactions on*, 56(9):1390–1393, 2008.

[WUB13] WD Wellisch, IA Ulian, and AN Barreto. Iterative correction of clipped OFDM signals with unknown clipping levels. In *Vehicular Technology Conference (VTC 2013-Spring)*, number 3, 2013.

[YW06] Guosen Yue and Xiaodong Wang. A hybrid PAPR reduction scheme for coded OFDM. *IEEE Transactions on Wireless Communications*, 5(10):2712–2722, October 2006.

[YYP00] Jaehyok Yi, Youngoo Yang, and Myungkyu Park. Analog predistortion linearizer for high-power RF amplifiers. *IEEE Transactions on Microwave Theory and Techniques*, 48(12):2709–2713, 2000.

[ZF05] P. Zillmann and G.P. Fettweis. On the Capacity of Multicarrier Transmission over Nonlinear Channels. *2005 IEEE 61st Vehicular Technology Conference*, 2(1):1148–1152, 2005.

[Zil07] Peter Zillmann. *Empfangsalgorithmen für nichtlinear verzerrte Mehrträgersignale*. Phd thesis, Technische Universität Dresden, 2007.

[ZQDR05] GT Zhou, Hua Qian, Lei Ding, and Raviv Raich. On the baseband representation of a bandpass nonlinearity. *IEEE Transactions on Signal Processing*, 53(8 Part 1):2953–2957, 2005.

LEBENSLAUF

PERSÖNLICHE ANGABEN

Name: Jan Dohl

Familienstand: ledig

Staatsangehörigkeit: deutsch

Geburtsdatum: 07.06.1983

Geburtsort: Mittweida

BERUFSERFAPRAXISHRUNG & PRAKTIKA

seit 10/2013	Gründungsmitglied “Airrays GmbH” Entwicklung intelligenter Antennensysteme
10/2009 – 09/2014	Wissenschaftlicher Mitarbeiter, Vodafone Chair Mobile Communication Systems, Technische Universität Dresden
04/2008 – 08/2008	Fachpraktikum, MAFI Transport-Systeme GmbH Entwicklung eines Fernwartungssystems für Industrieanlagen
10/2003 – 03/2008	Werkstudent, MAFI Transport-Systeme GmbH Hardwarenahe Softwareentwicklung
04/2005 – 06/2005	Praktikum Softwaretechnologie, T-Systems MMS Dresden Entwicklung eines Systems für Online-Pferdewetten
04/2003 – 09/2003	Praktikum, MAFI Transport-Systeme GmbH Entwicklung grafischer Bedienoberflächen für Sonderfahrzeuge

SCHULE & STUDIUM

10/2009 – 11/2014	Promotion: “Blind Estimation and Mitigation of Nonlinear Channels” Vodafone Stiftungslehrstuhl für Mobile Nachrichtensysteme ○ Abschlussnote: Magna Cum Laude
01/2009 – 07/2009	Diplomarbeit: “Charakterisierung von breitbandigen OFDM- und Single-Carrier-Systemen hinsichtlich der Sensitivität gegenüber analogen Hardware-Nichtidealitäten” Vodafone Stiftungslehrstuhl für Mobile Nachrichtensysteme ○ Abschlussnote: sehr gut (1,1)
10/2008 – 02/2009	Seminarreihe “Ready for the Job” ○ Schlüsselqualifikationen für das Arbeitsleben ○ Abschluss mit “sehr gut”
10/2003 – 07/2009	Studium der Informationssystemtechnik, TU Dresden ○ Schwerpunkt Elektrotechnik: Nachrichtentechnik ○ Schwerpunkt Informatik: Verteilte Systeme ○ Abschlussnote: sehr gut (1,3)
07/2002 – 04/2003	Zivildienst, Mittweidaer Werkstätten der Lebenshilfe e.V. Betreuung, Fahrdienste
08/1994 – 06/2002	Abitur, Gymnasium am Schwanenteich Mittweida Abschlussnote: 1,5

Jan Dohl • Südhöhe 33 • 01217 Dresden
Telefon: 0151-59211238 • E-Mail: jan@legxaro.de

EHRENAMT, AUSZEICHNUNGEN & FREIZEIT

11/2009	Hermann-Willkomm-Preis für beste Diplomarbeit im Bereich Informationssystemtechnik 2009
07/2009	Dankesschreiben des Rektors der TU-Dresden für ehrenamtliches Engagement in der AG DSN
05/2007	Sieg beim Rhode & Schwarz Fallstudienwettbewerb "Get the Signals" an der TU Dresden
2006 – 2009	Systemadministrator (Root) der AG DSN
2006 – 2009	Vertretung der AG DSN an Linux Info Tagen Support von Teilnehmern bei Linux Install Partys
01/2006	Philips Preis für bestes Vordiplom des Jahrgangs IST/03
2005 – 2009	ehrenamtlicher Nutzerbetreuer & Entwickler Arbeitsgemeinschaft Dresdner Studentennetz (AG DSN) Sektion Wundtstraße
Hobbys	Fotografie, Radfahren, Wandern, Tanzen

FACHLICHE KENNTNISSE

Kenntnisgrade: + Grundkenntnisse, ++ fortgeschritten, +++ sehr gut

Gebiet	*Teilgebiet*	*Kenntnisgrad*
Betriebssysteme	GNU/Linux	+++
	Windows (XP/7)	++
Programmiersprachen	C/C++	+++
	Matlab/Simulink	+++
	C#.NET	++
	Java	+
	Python	+++
	SQL	++
Anwendungen	MS Office	++
	MS Project	+
	OpenOffice	++
	LaTeX	+++
Sprachen	Deutsch	Muttersprache
	Englisch	sehr gut in Wort und Schrift

Dresden, 1. Juli 2015